ELOHIM

A New Perspective of God, Science, and Creation

L. Fernandez

Fiesta
PUBLISHING

Elohim: A New Perspective of God, Science, and Creation

ISBN: 979-8-9901765-0-8 (Paperback)
ISBN: 979-8-9901765-1-5 (eBook)

Library of Congress Control Number: 2024906050

Author: L. Fernandez
Front cover image by: Creative Instincts
Interior book design by: Creative Instincts

First printing edition 2026

Printed in the United States of America.

10 9 8 7 6 5 4 3 2 1

Fiesta Publishing
PO Box 44984
Phoenix, AZ85064

www.fiestapublishing.com

Dedication

This book is dedicated to Elohim who inspired and assisted me in its creation. In it He reaches out to His favorite creation, Homo sapiens sapiens. He hopes to stir some primeval memories of His interacting with the populations two million years ago, as they evolved. As He watches their current actions, leading to instability in so much of His beautiful earth, He thinks sadly of the evolution of all creatures and the loving relationship He had with them. As early Homo Sapiens evolved, less than two hundred thousand years ago, all intermediate species had the compassionate loving qualities of their Maker. Even the early days of Homo sapiens were, as far as we know, peaceful. With the creation of man to His ultimate level of Homo sapiens sapiens about two hundred thousand years ago, the harmony of life and death as He intended for each species He created had changed. He still confirms His love for them and all His creations and wishes that they re-examine their relationships with each other and with the earth at large. This book only goes as far as the early establishment of Homo sapiens sapiens as the sole human species to survive and become the main sentient occupant of the earth. The roots of the greed and destruction that followed as He advanced through the centuries was not yet manifest then. The book ends with Elohim's sad conclusion that the future would be turbulent as Homo sapiens spread across the earth. I had planned for a continuation of man's historical exploits but felt little enthusiasm for it as I saw what damage was done through the centuries.

Hope still exists that human beings will free themselves from the unrest that exists in many geographical areas. All they have to do is individually refuse Satan's help and instead ask Elohim for help. That as a lifetime policy can help us solve many worldly problems. Once we get used to doing this, even individual lives can be transformed.

Humans arose, rather, as a fortuitous and contingent outcome of thousands of linked events, any one of which could have occurred differently and sent history on an alternative pathway that would not have led to consciousness. To cite four:

1. *If our inconspicuous and fragile lineage had not been among the few survivors of the initial radiation of multicellular life in the Cambrian explosion 530 million years ago, then no vertebrates would have inhabited the earth. [The only fossilized survivor of the initial cordate phylum is a tiny creature called Pikaia.]*

2. *If a small and unpromising group of lobe-finned fish had not evolved fin bones with a strong central axis capable of bearing weight on land, vertebrates would have never become terrestrial.*

3. *If a large terrestrial body had not struck the earth 65 million years ago then dinosaurs would still be dominant and mammals insignificant.*

4. *If a small lineage of primates had not evolved upright posture on the drying African savannas just 203 million years ago, then our ancestry might have ended in a line of apes.*

Stephen Jay Gould
Scientific American, October 1994

Table of Contents

Prologue

We are in an era of explosive new discoveries in every field of science. For most of us, the fine points of this constant advance of the learning curve are beyond comprehension. For the most part, we focus only on areas where that curve may impact practically on our lives. In doing so we fail to realize that all scientific knowledge is a spectrum that moves from established fact to a variety of most likely narratives, the largest part being in the latter category. This applies particularly to the field of paleoanthropology, the science of how living things came to be and advanced to where we are today. Common belief systems subscribing to how we came to exist vary from a narrative claiming sudden creation by a creator about six thousand years ago, to the Darwinian theory of evolution and the origin of Homo sapiens sapiens about eighty thousand years ago. The science of how this occurred is equally confusing. Darwin's theory of evolution gave rise to the newer modern synthesis and was further questioned by the theory of punctuated equilibrium put forward by Stephen Jay Gould and Miles Eldrich. All of evolution, up to the appearance of Homo sapiens sapiens occurred in the prehistoric era, well before the invention of writing. The first written documents occurred about thirty-six hundred years ago. They were found in Syria in an area occupied by a people who called themselves Sumerians. Prior to that time all knowledge was preserved by memory of people recounting important events verbally to each generation. This is not an event preservation method that produced accuracy. Our understanding of what happened before that time is dependent on the study of fossils and preserved structural remains and tools. A well-developed system for assessing the age of these artifacts allows archeologists to reach conclusions about the habitants of a particular time were like. While timing can be quite accurate, the conclusions reached from examining artifacts can vary.

Thus, many crucial areas of knowledge are very controversial. Today the most accurate assumptions come from genetic studies of extant relatives of the fossilized specimens. Unfortunately, in several key areas, fossils are scarce and are often found only in the form of individual portions of skeletal structures. This bulk of evidence, and the conclusions of different theories lead to the knowledge that we have today. It varies from largely incomplete or non-existent fossil trails and a variety of hypotheses. Certain milestones in the march from single cell life to multicellular life, and from there to the major kingdoms of life, are well established by an authenticated fossil record. More recently, advances in determining the accurate aging of fossils are possible by adding information from analysis of genes; the how and why is still mostly based on plausible narratives. To these many scientific and metaphysical narratives that war with each other on how man came to be, I choose to add my own narrative, a combination of the two.

To talk of a creator or an all-encompassing God has become very unfashionable in our modern world. We are so awed by what science and technology can give us that the very concept of "God" seems tame in comparison. But what if all of this amazing science is an inherent part of what God is all about? It does not take a stretch of imagination to believe that science is simply our incompletely understood conceptualization of the Creator's method. In this version of Him, Elohim is a God who has experienced many previous creations. Most have ended with the destruction of worlds by the very beings He created (note that at this point I am inserting my own imagined story of God the Creator). Much of the paleontological info we have today is filled with gaps in knowledge when looking for explanations of how everything happened and how long it took. Even though you could accuse me of making everything up, I must point out that when we don't have reliable scientific facts, there is always a theory or a hypothesis that gradually gets sponsored by well-known scientists and eventually is accepted as "close" to the truth. True experimental proof is difficult to produce for past events and living things which are now extinct. That is why a creationist's perspective is not so different from a scientist's. We are all guessing about the past. We can never actually picture what it was like. That changed thirty-six hundred years ago.

The event that occurred was the invention of writing. Things took a big leap forward when flat stones with writing scratched on them were found. Those were the first written words. Writing soon became the prominent vehicle for communication. Following this event was verbal transmission of history among subsequent generations. Over time, recorded written descriptions and accounts of experiences became the source of historical knowledge and with written descriptions, we dropped imagined history (not a criticism of history). The institution of the written word changed industry and society drastically since all experiences documented were recorded in a largely homogenous way. As methods of dispersing the written word improved, permanent books that preserved knowledge spread through communities transforming industry and social interaction among distant populations.

Scientific inquiry became a reality and important findings were recorded and preserved since the third millennium. Later, when paper was used for writing, it became the main mechanism by which knowledge was made available to ordinary people. New information was disseminated more efficiently and was saved, while rapid changes in dispersion of manufactured goods occurred.

As man's environment improved which allowed him to simplify the essential processes of obtaining food and protecting himself from environmental nature, Elohim, who was helping them along, received the appropriate credit for their improved lifestyle. He has many different names. As Elohim interacted with all His early creations, they knew Him and had a loving relationship with Him. This interaction continued until Homo sapiens became the most dominant creature. Their early relationship was similar to those of other animals and He communicated with all the species of the genus Homo. They were comfortable around Him when He made Himself visible.

I paint with a wide brush an entity that has experienced many previous creations of universes and over unimaginable amounts of time has with exquisite detail produced the current universe, a thing of beauty in its great symmetry and logic. Vast amounts of scientific data exist on the state of our universe from its origin to its current state. This scientific narrative develops as scientists analyze the world we live in with special telescopes in space. Though we have unearthed many wonders, there is still a lot for us to visualize and analyze. Despite scientific findings, there is much that remains to be uncovered, new techniques for visualizing, sampling, and measuring all that is around us and in outer space.

The story of how we came to be, even when backed by massive amounts of scientific data, is incomplete and has gaps. These gaps which have no scientific explanation are current mysteries in the process of evolution. These are gaps that I put forward as evidence that the hand of God exists even amidst all the science. Most of the gaps disappear when a divine entity credibly is held responsible for them.

The origin of the universe is currently explained by the theory of the Big Bang. The initial bang was followed by a rapid inflationary process giving rise to the formation of galaxies with their billions of stars and solar systems. In the initial throes of this process lies the first mystery of our existence, the dominance of matter over antimatter. If things had slanted the other way our material world would not exist or may be a radically different reality. The result of the annihilation of most antimatter leaving only a small amount of matter is still a mystery. Why had all antimatter been eliminated and by whom? While we can only guess at the *why*, the *who* can easily be a construed to be a powerful unknown creative deity—God. He engineered it so that only a small amount of matter was left, and almost all of the antimatter had been destroyed.

Having equal amounts of both would have made it impossible for life as it is to develop. The progression of life from the first amino acid to RNA and DNA could not have evolved in the absence of the ingredients that Elohim provided for the creation of His first cell. It could not have occurred naturally because it needed all of the essential ingredients to be available simultaneously and none of them even existed as yet. It was an impossibility without Elohim helping it along and producing ingredients as they were required. For any of the scientific phenomena described, this would not occur in the natural process of evolution. Nor would there be any of the essentials needed for life to promote growth and differentiation. For the same reason even the first cell would have trouble maintaining life if it just happened by chance.

Elohim, not the most well-known name by which we address God, is presented here as a divinity that transformed from a cold silent singularity to our universe and conformed the matter that formed the singularity into the world we live in. The participation of a Creator in the fabrication of the universe does not negate scientific knowledge. Much of science simply portrays human understanding of the Creator's process. Science and God are therefore not incompatible. The reader may read this as a simplified chronological scientific account of the origin of the universe and the evolutionary process. One should be aware that our scientific knowledge is incomplete and that there are many mysteries that still remain unanswered. Some I outlined above. More will be discussed in my narrative. Many of these are easily explained if one assumes the existence of an extremely intelligent entity that carved out our world with great attention to detail. He organized it for the specific benefit of the living things that He would populate the earth with.

Whether you are a believer, an agnostic, or an atheist, it never hurts to look at a great mystery from a different perspective. At the least, I hope you enjoy the simplified narrative of how we came to be. If you can look at intricate diagrams of the atom and its particles, all done up in color and accept the written description of it, you would never guess that they may be nothing like the drawings. They are imagined on the basis of what scientists have discerned to be the components of the atom and its constituents. It is not solid. It is mostly a space, at the core of which sits the proton and the neutron. The proton has a positive charge and the neutron no charge. Around the core are orbiting electrons which form orbit-like passes throughout the empty space. They carry a weak positive charge. The orbits of the electrons are arranged so that they form concentric but not parallel paths progressively inward from the outermost virtual shell of the atom. That is my verbal description of the diagrams that I have seen in texts on the subject. I know that I am looking at an atom and that the pictures are as close as I can get to what the atom really looks like. I take them as they are offered, and the atom becomes a reality to me. Similarly, belief in a God I have not seen but can see the works of, is enough evidence for me.

The investigative works of several scientists, one building on the another's work, created the entity that is the atom. With J. J. Thomson and Robert Millikan producing significant contributions to charting the three-dimensional appearance, the structure of the atom became real. When George Francis FitzGerald suggested that the atom might be similar to a tiny solar system, a derived image began to form. Niels Bohr confirmed its reality when he created entirely new mathematical equations that worked when challenged by reality. They worked in predicting the behavior of matter. The physical and chemical structure of the atom was constructed and eventually imagined based on all the various energies detected by many different experimental scenarios. Elaborate diagrams of each atom in the periodic table were created. Another offshoot in this investigation was quantum mechanics. This was described by Niels Bohr, and he found a whole new type of mathematical equations to describe the microanatomy and energy-producing components of the structure of matter. Their success is evident in the final picture they achieved. The chemical equations, the physical properties, spectroscopic appearance, and nuclear radiations each produced equations that fitted together. There was a practical impact! Based on the findings, the cathode ray tube became the heart of two new industries—computers and television sets. Many more new industries have evolved since then.

This history describes some amazing work from humans obtained by collusion between investigators. Elohim was very proud of them. He had given help to any who asked for it. Eventually man would discover more and more of the workings of the universe He had created and hopefully produce a world that provided humanity with constantly improving quality of life.

So, my version of the progress of evolution presents the science as we know it, but gives the credit for unexpected persistence of the fragile beginnings of life to Elohim because there was no sentient life on earth at that time that could have done it. I have greatly simplified the science to general concepts and key event milestones.

Chapter One

The Beginning

13.8 gya

Elohim stirred out of His deep sleep. Images of pain and rejection fought against a need to discover and create. Slowly the need to exist won the day and He let himself sense the world around Him. Dark, dense matter surrounded Him, condensed to a density vastly greater than that of black holes observed within our galaxy. He had existed this way for eons, dormant in dreams and memories. As He awoke, vague memories of light and blue skies emerged. Were there other memories? Of love and rejection, creation gone astray, and of a terrible urge to end it all? The memories vanished. They were too painful, even for a being that knew no limits to His power and awareness. Elohim was like an infant born into a new world with genetic memories of generations gone by but unable to recall the details. The urge *to be* grew gradually and became urgent, for that was in His nature. Flashes of memories of previous creations, each ending in annihilation, were quickly replaced by ways to improve in His new venture. This time He would achieve His goal of bringing about the ultimate creation. His new sentient species would have all the inherent spiritual, ethical, and intellectual qualities inherent in Him, as did previous creatures.

Vividly He saw all the sentient creatures of past universes. Each was physically different and was optimally adapted to the environment of the worlds He had created for them. Physical aesthetics meant nothing to Him, for the beauty and harmony of the spirit, or soul, was His ideal. This time He would create a creature that would overcome the distractions of the physical world around Him and remain loyal to His spiritual aesthetic. Elohim decided that love, the most powerful of all His energy forms, would influence the creature's actions. Man, the creature that He visualized as His next sentient being, would be created with the imperative to promote peace and harmony among all living things.

Elohim assessed the physical status of His fabric. Immense gravitational attraction held matter frozen in a tiny volume. The energy waves of matter were lying dormant within this gravitational field in the form of what we today conceive as a unique and powerful form of a Higgs field. Having gone through this process with countless other universes, Elohim knew exactly what to do. In an instant He initiated the process of activating this field. Huge repulsive forces were released resulting in an instantaneous repulsion of all microparticles of matter and their waveforms. This occurred at zero to 10-43 seconds. The inflaton suddenly underwent rapid inflation in volume. In less than 10-35 seconds the matter in the singularity expanded to

1030 times its volume. Rapid heating of the energized particles was followed by rapid cooling as they were separated from each other by inflation. These numbers are impossible to imagine. A comparative scenario has been visualized as a DNA molecule being scaled to the size of our Milky Way galaxy in an interval that is less than a billionth of a billionth of a billionth of a blink of the eye, by Brian Greene, a well-known astrophysicist.

The initial expansion of particulate matter was chaotic. Elohim wanted to set in motion a process that would minimize the tendency for energy systems to deteriorate into chaos. He desired a long lifespan for His universe. During and following the initial rapid inflation described above He introduced a more orderly outward movement of particulate matter causing it to be distributed uniformly throughout its volume. Its projected density in the initial stages had to be very close to a density called the critical density. If it were higher, it would have collapsed back to singularity. If it were lower, it would not have formed galaxies. The features of this critical density are very exact. Any variation of the initial density by more than one part in 1055 would result in either collapse or rarefication of the new universe. Furthermore, scientists believe that the chance of the initial density happening on its own was very small. It was not a problem for Elohim. He could manipulate matter as He chose. He knew all its capabilities and susceptibilities.

We will see this kind of serendipity recur frequently as we move through the progress of our universe's formation. For instance, scientists project that at the temperatures of the early Big Bang there should have been large numbers of exotic elemental particles formed. Even if some had decayed with time, there should have been relic particles as evidence of their existence early on. None have yet been found. Because of these and other discrepancies, the Big Bang theory has been adjusted to what is called *inflationary cosmology.* None of the many theories, however, explain why the initial reaction occurred or from where it came. Although science has come a long way in its understanding of how energy and mass function and has done an even greater job applying this knowledge to practical uses, there is still much it does not know about how the universe came to be and how it currently functions.

It should be mentioned that the universe did not start with the bang. There was an undefined interval in space time that occurred before the inflaton began. Its density, by definition, would have to be greater than 1093 g/cm. We call this situation the Planck density. It is an area about which we know very little. Scientists know that quantum physics does not work within it. They are not even able to produce such a space. These two facts have led to the concept of the Big Bang and the theory that followed as not being adequate. Instead, scientists are focused on the initial inflaton followed by controlled inflation that took over after the activation of the singularity. They have been able to assign workable physics to the inflaton process and produce probable sequelae that occurred once the initial rate of inflation slowed down. We know from their studies that the universe has continued to inflate since that time and that it is starting to accelerate slightly.

By 10-6 seconds after the onset of the inflaton, elementary particles of matter were so separated in space that the energy of the force of gravity and the Higgs field were reduced, leading to a fall in temperature. This led to the combination of elementary particles of matter to form protons and neutrons. At this early stage of the universe all particles of matter were accompanied by an equal number of antimatter particles. Elohim adjusted the numbers of protons and neutrons to be slightly more than the number of anti-protons and anti-neutrons. When the temperatures dropped low enough, the matter-antimatter pairs interacted and destroyed each other, leaving intact a small excess of protons and neutrons and electrons. All the anti-protons, anti-neutrons, and anti-electrons were eliminated. A few minutes later as temperatures were further reduced, atomic nuclei, mostly deuterium and hydrogen, began to form. It took another 369,000 years for the nuclei to combine with electrons and form atoms. During this interval the universe remained dark. At the end of 369,000 years atoms were formed, and this resulted in the release of electro-magnetic energy waves including photons. When this happened, the photons traveled away from the source accompanied by other electro-magnetic radiation that was released introducing light to the universe. The preceding 369,000 years were labeled the Dark Ages.

This initial radiation continued and is called cosmic radiation. It is detectable as cosmic microwave or infrared radiation by spectroscopic telescopes orbiting earth in space. Measurement of this cosmic radiation shows it to be distributed evenly throughout that portion of the universe that we perceive. This generalized homogeneity was interspersed with areas where matter was collected in small dense pockets. As the density increased the hydrogen in the dense areas became activated and the temperature increased. Eventually the heat and pressure were so great that nucleosynthesis of helium from hydrogen occurred. These areas developed into suns and over centuries collections of dust and debris called nebulas formed and are surrounded by an active sun-producing area. As these clumps of matter drifted away from each other due to continued inflation, galaxies formed.

These were just larger versions of the nebulas and sun-producing areas. Elohim looked forward to watching this process with considerable anticipation. In His experience no two universes had been alike.

This new universe would have its own unique composition. Only 5 percent of the mass of the universe was baryonic matter and formed galaxies, dust clouds, and the occasional single asteroid or comet in outer space. When living things evolved in the distant future, they too would be a part of this baryonic mass, called gravitational mass. The remaining 95 percent was divided between dark matter and a mysterious force called dark energy. Dark energy or matter constituted 27 percent and dark energy 68 percent of the mass of the universe.

Elohim had an important role for dark matter. If young galaxies collected in a group were left alone, gravitational force between them would pull them closer to each other until they collided with one another, eventually forming a black hole. By providing a matrix of dark matter that infiltrated the space between various bodies within a galaxy, He provided a repulsive force equivalent to the attractive force called gravity that baryonic matter exhibited. This would keep the relative positions of all objects stable. His next step was to configure the space between the galaxies that had slowly increased because of the continued expansion of the original inflaton.

To control this, Elohim introduced dark energy. This mysterious force filled the gaps between galaxies. It is also present within galaxies and perfuses all baryonic matter. Elohim controlled its rate of expansion to optimize the growth of the universe according to His plans. He planned to produce expansion of space by steadily increasing the dark energy. By increasing the distance between existing galaxies and possibly beyond them, this substance became *space itself.* It is of interest that this expansion is accelerating. Although we cannot feel it as we do baryonic mass (gravitational pull), we know that it is an unknown type of energy that carries mass effect. So far, all attempts to create it have failed. When space is tested for baryonic matter, it is found to be close to a vacuum with a small amount of hydrogen gas floating around. Dark energy does not obstruct the movement of baryonic mass, such as planets, asteroids, or comets as they move through it. On the other hand, as it expands it pushes galaxies and nebulas further away from each other by putting more dark energy between them. This is true even though there is evidence that dark energy passes unnoticed through and perfuses all baryonic matter. It is one of the least understood mysteries of our universe.

Elohim had a special purpose for His dark energy. It would be His medium for communication with each living thing in His universe. Since it occupied all of space and would perfuse and penetrate through all living things it was a small matter for Him to set up a subliminal communication system with living beings in the future. It would also be the pulse on the entire universe keeping Him aware of everything going on in it.

Elohim looked far into the future and caught a glimpse of two men who were trying to predict the behavior of matter in terms of mathematical formulas. In 1664, Isaac Newton was sitting and pondering the possibility of introducing gravitational energy into his equations for the three laws of motion and inertia. He had already discovered and introduced the laws of motion which were being put to good use in the field of mechanics. His laws of motion were:

First Law: A body remains at rest or in motion in a straight line at a constant speed unless it is acted upon by a force. This is called the Law of Inertia.

Second Law: The rate of change of momentum of a body is equal in both magnitude and direction to the force imposed upon it. The momentum of a body is equal to the product of this mass and its velocity. This statement contains several equations that can be derived to display various relationships of the vectors involved. For a body whose mass "m" is constant it can be expressed with:

> F = ma, where "F" is force, "m" is mass, and "a" is acceleration. This means that if a body has a force acting against it, it is accelerated. If there is no force, it is not accelerated.

Third Law: When two bodies interact with each other they apply forces to each other that are equal and opposite. If this is the situation, the two forces are said to be in equilibrium. No acceleration occurs. Interestingly this could apply to a book sitting on a table. The book pushes down, and the table pushes back so no acceleration occurs. It is a different sort of example of forces in equilibrium.

Newton was sure that gravity needed to be inserted into the calculations since it was a universal force that caused attraction between bodies of matter. This was when he asked Elohim for help. Soon after that he came up with a solution and was able to define the laws of gravity in relation to motion.

Elohim guided Newton's thoughts every now and then to set him on the right path. He knew that it would take just a little push in the right direction to get a man as brilliant as Newton to see the light. He was right. Newton went on to establish laws that described how matter behaves as a function of its mass and how mass and movement combined give estimates of energy usage in measurable units. He outlined several equations that described the behaviors of objects in motion. This established the science of mechanics. Isaac Newton was responsible for advances in many fields including those of mathematics, physics, and philosophy. He was a firm believer in a God who was the Creator of our world and felt that the Creator continued to intervene and guide events in the universe.

Almost three hundred forty years later, Albert Einstein, a theoretical physicist, began working on movements of objects in space. He came up with a new set of rules. He defined a new relationship between space, time, and mass when looking at movement in space. He called it special relativity. We have learned over time via Newton and Einstein that there are two different sets of mathematics to explain the behavior of baryonic matter. Newton's formulas mostly address the behavior of objects on earth where gravity is the prime mover. Einstein, on the other hand, went on to examine the behavior of larger bodies like planets and galaxies in space and he found that things work differently in that setting. Large bodies like planets do not move in a straight line but have a slightly curved trajectory. Also, the perception of reality when applied to the movement of large bodies like planets and galaxies is often misleading. Both the distance and the time it took were hard to measure accurately. For example, two observers who were stationary looking at an object moving from one point to another found that they had slightly different measurements of distance moved and time taken. We are unable to synchronize them accurately enough. He also determined that the energy used for movement in space was influenced by a space-time relationship instead of time and space as independent variables. In essence there was no absolute space and no absolute time but there was absolute spacetime when looking at movements of objects in space. It was difficult to set up a suitable experiment to prove his theory.

After establishing the theory of special relativity, which describes the relationship between time and space and movement in relation to the speed of light and other energy forms, Einstein tackled the concept that bothered him the most—he did not feel that aether played a significant role in the movement of mass in space. According to established Newtonian physics at that time, movement was an absolute entity which was affected by travel through the *light-carrying aether*—that was space. Einstein felt, when he looked at this in thought experiments, that there was a sparsity of evidence for the reality of aether. To him it did not feel like a part of the natural process. He did find that measurements of duration and simultaneity of an event in time were related to the observers being stationary. In such a situation, the speed of light (and other electromagnetic waves) is constant. When a unit of mass started moving from point A to point B, if you had two stationary observers record the time it took and the distance it made, you would find each recorded different values for both time and distance traveled. On the other hand, velocity was always the speed of light (and other electromagnetic radiation), which was constant. He implied that in mathematical equations *space and time* somehow adjusted their measurement values to produce a value of 186,000 miles per second, the value of the velocity of light. They appeared to be one entity which he called *spacetime*.

Isaac Newton laid down definitions for absolute time and absolute space, both of which existed without any external influences. Einstein decided he had to discard these and find new equations to represent the phenomenon of the relativity of time and space. He published his perception of special relativity in a paper in June 1905. It stated that Newtonian laws of physics would apply in stellar systems if inertial frames for all objects were uniform. The latter statement implies that, if tested by an accelerator moving within the system, the test would detect no acceleration in any of the frames. Secondly, he postulated that the speed of light (or any other electromagnetic radiation) in a vacuum is always the same and nothing is faster. He also stated that the speed of light would be the same for all observers no matter how fast they were traveling.

While Einstein was conjuring with his thoughts on the explanation for the behavior of light waves and the relationship of the accepted concept of a substance called the aether, a review in scientific papers related to the aether caught his eye. Michelson and Morley published an unconnected paper showing no evidence of any aether when doing experiments with light beams in space. Several other scientists corroborated this idea. Another advance that worked its way into his speculations was the discovery by Maxwell in 1894 that electricity could be converted to magnetism and vice versa. During his experiments, he noted that both forms of energy behaved similarly as they traveled from one point to another. Einstein felt his responsibility as a theoretical physicist was to put this all together. He came up with two postulates. The first he called the *principle of relativity,* which holds that "the laws of mechanics and electrodynamics are the same in all reference systems moving at constant velocity relative to one another." He also went on to definitively state "that the introduction of *light ether* will prove to be superfluous. . . ." By this simple statement he made all kinds of things plausible.

He even addressed time dilation. Take a set of twins—one male, one female. If the man stayed on the platform and the female traveled at *almost* the speed of light a long distance and returned, she would seem younger to the male. But because motion is relative, it would be hard to say who moved from who. To the female twin it would appear that the male was moving away from her and when he returned (to her perception) he would appear younger to her. It was all relative. This phenomenon called time dilation has been immortalized as the twin paradox. All this discovery just by sitting and thinking hard about it.

Einstein went one step further. He addressed and published his theory of general relativity in 1915, ten years after the publication of special relativity. This proved to be a more complex process. He collaborated with multiple scientists combining their practical findings with his theoretical ones. It bothered Einstein that his relative theory only applied to objects in motion. He felt that he needed to establish a similar theory for objects in motion. His instinct was that it was somehow related to gravity. His theorizing told him that a person in a closed, windowless container standing on one surface of the cube with his feet firmly planted, would not know if it was because he was in space with a steady force pushing his container upwards or if the container was at rest in a gravitational field like a stationary surface on earth. From this thought he introduced his *equivalence principle*. He felt he could use this principle to apply relativity to moving systems. It was proven that inertial mass and gravitational mass are equivalent. Perhaps all inertial effects, such as resistance to acceleration, and gravitational effects, such as weight, were equivalent. He thought about it and saw relevance in the term *inertia-gravitational field.* He decided he could use the occurrence of a light beam being bent by forces to set up an experiment. This was his thought experiment.

A chamber is being accelerated upward. A laser beam is directed through a pinhole on one wall. By the time the beam hits the opposite wall it should strike lower on the opposite wall. It would travel a curved path to this spot. The equivalence principle states that this effect should occur regardless of whether the test occurs in space with an accelerated container or if it occurs on earth with a stationary container.

Einstein did not follow through on this resolve for four years. Then, in 1911, he decided to try again to get a theory of relativity in moving objects or a general theory of relativity. He did a series of calculations based on thought experiments and came up with a measurement using the gravitational field of the sun. He predicted that a ray of light passing the sun would undergo a deflection of 0.83 arc seconds. He immediately thought of a workable scenario. In an eclipse of the sun, the stars and planets that are close to the sun are visible. The problem was that he needed special mathematics that addressed the physical situation of an area close to the sun. He went to his friend Marcel Grossmann, a mathematician, for help. Grossmann introduced Einstein to the mathematical concept of non-Euclidian geometry, called tensor calculus, introduced by Bernhard Riemann. He was a child prodigy who was working on the geometry of curved surfaces with Carl Friedrich Gauss, a German mathematician and astronomer.

For the next year Einstein pursued the optimal math for an experiment around the curved surface of the sun. He finally had an idea. So far, he had only focused on artificially derived experiments. What if he looked for a natural situation that fit the requirements? To make a long story short, he thought of one. He imagined, what if he had his astronomer friends set up an experiment to be conducted during an upcoming total eclipse of the sun that would take place on August 21, 1914. They made their plans, but unfortunately, World War I started twelve days before the eclipse date. It was for the best because Einstein found glitches in what he called his *Entwurf* theory. For a year after this Einstein worked on establishing the right equations to describe the bending of light waves when passing a much bigger object in space. Finally, in 1915, he found the right mathematics to do this. To confirm his theory, he worked with astronomers to test it as Mercury passed the sun in perihelion, the shortest distance from the sun. In 1916, he published the Einstein equation which defined the effects of gravity when a small body passes a much larger one in space. The equation described the bending of light waves by the gravitation of a massive body. It introduces a four-dimensional reality—matter, space, time, and energy. John Wheeler, a physicist, describes it as, "Matter tells spacetime how to curve and curved space tells matter how to move."

Einstein needed a formula to enable a workable model for predicting the behavior of planets and galaxies in space-time units with gravitational force incorporated into the formulas. He tried to create formulas to define what happens to mass and energy when two light pulses are projected in opposite directions. He sat and thought about the content of related papers produced by many other scientists and eventually concluded that the force of gravity remained a mysterious entity. His conclusion: "The mass of a body is a measure of its energy content." His theorem for this was, "If a body emits the energy L in the form of radiation, and is moving at a velocity V, its mass decreases by L/V2 or L=mV2." He later changed this to lettering E=mc2 with "E" for energy, "m" for mass, and "c" for the speed of light, which is fixed. So, "L" and "m" are the only variables. And it worked for his calculations!

Both Newton and Einstein dealt with matter as an experience we see and feel. It was easy to imagine the connection of various forces. Then, in the 1900s, Max Planck introduced mathematics that would describe the behavior of elemental particles or waves of energy. He introduced his quantum mechanics, which deals with the behavior of elemental particles like quarks, gluons, and electrons. This introduced a field of physics and mathematics that deals with invisible actors and needs a new system of mathematical formulas to define events in both space and on earth. Planck, a religious man, believed in a non-personal God as Creator. His quantum mechanics did not quite fit in with Einstein's theorems of relativity. Einstein tried for the rest of his life to define a unified theory connecting conventional physics with quantum physics. He was unsuccessful. The search continues. For most of his life Einstein believed an intelligent creative entity was responsible for the creation and running of our universe. The order and beauty of the so-called works of nature were too intelligently planned to have evolved independently. He referred to the *"old man"* as a divine entity frequently until the end of his days. He loved to think he was unraveling the mysteries of physics that had been laid down by Him. Some of his comments to friends and colleagues illustrate a genuine sense of controlling intelligence.

In his autobiography of Einstein, Walter Isaacson talks of Einstein's comments on the God he perceives. In one of these conversations he talked about his dissatisfaction with the fact that no causal mechanism had been found to explain the laws of quantum physics. He was a strong believer in looking for and finding causality before going ahead with adopting a new science. He talked about this with Isaacson. He stated the belief of Baruch Spinoza, his favorite religious philosopher, who lived in the seventeenth century, on this. Spinoza talked of a need to probe the causal relationship of all phenomena. Einstein added, "Especially at a time when the success of efforts to achieve a knowledge of the causal relationship phenomena was still quite modest." He was talking of the need for this to be done for quantum physics. According to Isaacson, Einstein did not believe in a personal God who interacted with man. He and Spinosa "both believed that a divine design was reflected in the elegant laws that governed the way the universe worked."

Isaacson goes on to state that this was not merely an expression of faith, but it was a principle that Einstein elevated to the level of a postulate, one that guided him in his work.

In the epilogue to Einstein's biography, Walter Isaacson tells of Einstein's notions of a God-like entity. He gives several quotes by Einstein, referring to the beauty and order in nature. He is quoted as saying to a friend, "A spirit is manifest in the laws of the universe—a spirit vastly superior to that of man, and one in the face of which we with our modest powers must feel humble." To explain this feeling, he stated, "In this way the pursuit of science leads to a religious feeling of a special sort." When he was asked about miracles, he stated, "The fact that the cosmos is incomprehensible, that it follows laws, is worthy of awe. This is the defining quality of a God who reveals himself in the harmony of all that exists."

To the end, he did not accept that quantum physics was completely understood. Despite its math working for many processes, he felt there was more to it that man did not yet understand. Of the many thought experiments Einstein devised to prove his theory of relativity, was one involving moving trains. He offered another scenario: Imagine lightning strikes a train track in two different places—A and B—simultaneously. And suppose the bolts are seen by two observers, a man and a woman. One will be on a fast-moving train looking out of a window at the exact midway point between A and B and facing the embankment. The other observer is on the embankment and is exactly at the same distance halfway between A and B. Light from each strike would travel from opposite directions at its usual speed of 186 miles per second, to the midpoint where both observers would see the flashes. How do we define *simultaneous*?

There are three situations. If the train paused at the moment of the strike, both observers would see the flashes as simultaneous. However, if the train maintained its velocity there would be some movement, however small, toward point B on the right for the man on the embankment so that he will see the light from B before he sees the light from A. He will therefore say the flashes were not simultaneous. The right flash occurred first.

The opposite would be true for the woman. She would be moving with the train toward the strike on the left and would be just a bit closer to that flash when it occurs and would say the left flash occurred first. There is no way to correct this result by repetition. It will always be the same. Therefore, all we can do is give a relative result. Einstein saw that this interpretation of reality was best represented by the merged modality of *spacetime*. There could be no *absolute time* and no *absolute space*.

Our theories both about God and the nature of our universe change constantly. The search for the truth is a characteristic of humanity that Elohim approves of, and He tries intermittently to help us understand. He is willing to face that challenge again, but first He must create a suitable environment for His creatures, one that would enthrall them and keep them searching for answers. For along with the ability to love and be creative, they would inherit their Creator's intelligence and curiosity and would spend time trying to understand the nature of His creation. Today, science attempts to understand the workings of the universe. Elohim, ever willing to satisfy the curiosity of His creation, helps with inspired ideas that are periodically sent to special individuals. As a result, with each century, we uncover new insights. Some mysteries continue to bother us and among these are the nature of dark matter and dark energy. The wonders of His creation are far beyond our comprehension and will present endless new problems for us to unravel. They keep us challenged and involved in the great universe in which we live.

Another recently found phenomenon that is amazing to ponder is the phenomenon of electron entanglement. It has been shown definitively by experiments that an electron can, by unknown forces, be expelled from an atom and travel a trajectory through the cosmos, even from one galaxy to another. Despite the distance separating it from its original home, it has been shown that it remains connected by a mysterious force with its original atom.

There is still much we need to learn about Elohim's universe. Elohim is thrilled at how much Homo sapiens have discovered in such a short time. He looks forward with anticipation to the next major discovery and hopes His assistance will be requested.

As millions of years went by, Elohim began to long for interactions with other entities. He remembered entities He had created in past universes, ethereal like Himself but with intelligence and emotions. They provided Him with companionship as He pursued His process of creating sentient creatures and their environments. He decided it was time to do this again despite the memories of not so pleasant histories in the past.

Chapter Two

Heavenly Unrest

The first creatures Elohim created were ethereal beings without substance. They were simply fractions of His essence, each with a unique identity. They communicated silently with each other and with Elohim and were initially all very loyal to Him and loved Him intensely. Like Elohim, they possessed the qualities of intelligence, curiosity, and creativity. They were also imprinted with His perception of what was right and wrong. Unfortunately, they contained a small element of His alter ego, trapped forever within Elohim's matrix in an endless struggle for dominance. It was this duality within His substance that introduced an unknown future in each of the universes He had previously created and would continue to create. Despite the birth and collapse of hundreds of universes, neither Elohim nor this entity, which we will call Beelzebub, had been successful in obliterating the other.

Of the two, Elohim was undeniably the more powerful. He had the power to control and manipulate all the energy forms that were part of His substance and to create material worlds out of them. The force of evil within its milieu had only one quality that gave it an advantage. Its actions were uninfluenced by any perceived moral imperative. In fact, it thrived on opposing all behaviors considered to be part of the moral code espoused by Elohim. Where Elohim taught His creations that love, compassion, and humility were the path to peaceful existence, Beelzebub appealed to the baser qualities within their personalities, those of pride, greed, jealousy, and the lust for power. Elohim believed that happiness lay in the service and care of other beings. Beelzebub knew that most lesser beings could be distracted by various forms of activities pleasurable to their egos. Preferring these activities, they would ignore Elohim's imperatives and listen to Beelzebub.

Elohim hoped that His angels would not be corrupted by Beelzebub and that He would be able to control any moves of disloyalty before they went too far. He had many categories of angels around Him. They were divided into different hierarchies, and each had a gift that enabled it to accomplish a special function. They communicated with each other telepathically and kept each other and Elohim company. In the beginning they existed peacefully with no sign of evil thoughts or actions.

Lucifer was the first angel that Elohim created and endowed with a larger portion of Elohim's spirit than any of His later creations. Lucifer initially saw things the way Elohim viewed them and did his best to guide and teach the other angels as Elohim expected him to do. As time went on however, he became increasingly impressed with his abilities and began to see himself as being equal to Elohim. Soon the baser qualities that were part of his makeup began to take over and influence his perspective. Jealousy became an increasing aspect of his persona, at first of other angels and then Elohim. He plotted in secret on how he might overthrow Elohim and take over the universe from Him. Over millennia Lucifer transformed into a persona that represented all that was evil.

Originally energized by the burning light of Elohim, as time went on, Lucifer gradually experienced a darkness that struggled to control him. Encouraged by suggestions from the hidden Beelzebub, he had doubts about the need for allegiance to Elohim and imagined that he could be as great and powerful as Him if he wanted. Lucifer resisted this darkness within him but did not think to ask for Elohim's assistance in this. Elohim on His part was aware of Lucifer's dilemma, but He could not act without a request for help. He had given His angels control over their destinies and could not interfere. He hoped that the other loyal angels would help Lucifer fight his battle against the evil within him. With further prodding from Beelzebub, Lucifer was unable to resist his urges and began to incite the lesser angels to help him overthrow Elohim. Gradually he developed a following of his own among his fellow beings. He enjoyed his stature and watched and imitated Elohim, always longing to be more like his Creator. His feelings of love and loyalty were likewise fiercer than the rest and for a long time he was content. But gradually, as he watched his fellow creatures adore and obey Elohim in all things, other feelings kept surfacing. Jealousy of Elohim and His angels continued to be his strongest emotion.

Archangel Michael oversaw the legion of protective angels. Like Elohim, Michael was acutely aware of all the thoughts and emotions of his fellow angels. He saw that the majority existed as he did, driven by love. Their love for Elohim remained unshaken and their loyalty made them willingly committed to His will. They were rewarded with the ability to see the universe and each other with love, and to interest themselves in the wonders of the universe with curiosity and amazement. The latter kept them occupied endlessly, probing the mysteries that they saw and sharing their findings with their Lord.

More than any other angel, however, Michael felt a responsibility to protect Elohim and his fellow angels. Despite Lucifer's attempts to hide his feelings, Michael was aware of the unrest within him. He tried to communicate with him. "Lucifer, my beloved brother, why are you so unhappy? Like the rest of us you are engulfed in the love of our Lord and have much to keep you occupied. Yet you seem to rebel against His love and sow discord among our fellows. I freely admit to you that I too have had thoughts of what it would be like to do whatever I like, regardless of the wishes of our Lord. An evil entity lurks deep within our souls and we have to work hard at controlling it. It seeks to use us to destroy our Lord and the cosmos. Would you rather live in such a world surrounded by hate, violence, and destruction for eternity, than in the blissful state that we now enjoy with our Lord?"

Lucifer felt a resurfacing of the bliss he had enjoyed when first created and was aware that he was being manipulated by the evil within him. "Michael, you are blessed with the will to resist that which is deep within you. I wish I had your strength and clear vision, but by giving us free choice in how we choose to act, the Lord has left us susceptible to the evil within us. The evil spirit within me is stronger than most. I will strive to fight the evil within and become more like you." Michael saw that for the moment he meant it and hoped that he would prevail. He decided to concentrate on shoring up the allegiances of some of his fellow angels who had begun to feel as Lucifer did.

Elohim, aware of all but not directly influencing the actions of any of His angels, knew that the eternal conflict with His alter ego had begun. In sadness He saw the endless possibilities for chaos stretched far into time. Simultaneously He saw the possibilities that His humans would prevail over the temptations of evil within them and hoped for better outcomes. Had this very conflict not been what kept His previous universes eternally interesting? Although He did not relish the prospect of destroying yet another universe and receding back to His solitary existence as a timeless singularity, He considered the fact that this new universe was very young, and the possibilities of achieving a good outcome were endless. They would keep Him engaged in meaningful relationships with His creations for millennia. Elohim turned His attention to what was going on in the rest of the cosmos.

Our theories both about God and the nature of our universe change constantly. The search for the truth is a characteristic of humanity that Elohim approves of and tries intermittently to help us with. He is willing to face that challenge again, but first He must create a suitable environment for His creatures—one that would enthrall them and keep them searching for answers. For along with the ability to love and be creative, they would inherit their Creator's intelligence and curiosity and would spend time trying to understand the nature of His creation. Scientists today attempt to understand the workings of the universe. Elohim is always willing to satisfy the curiosity of His creation. He helps them with inspired ideas periodically that are given to special individuals with each century, who uncover new insights. Many mysteries however continue to bother us and among these are the nature of dark matter and dark energy. The wonders of His creation are still far beyond our comprehension and will forever present new mysteries for us to unravel.

Chapter Three

The Banishment of Lucifer

How art thou fallen from heaven,
O Lucifer, son of the morning!
How art thou cut down to the ground,
which didst weaken the nations!
Isaiah 14:12 (KJV)

The day came when Lucifer could not be ignored any more. The love Elohim implanted in him had been replaced almost entirely by jealousy, hatred, and an overwhelming need for power. The pain of rejection was a daily experience for Elohim as more of His creatures became like Lucifer and those that did not were subjected to violence and isolation by Lucifer's horde of followers. He knew that He needed to protect those that remained loyal. In an instant He created a partition in the universe and formed an alternate dimension in which the angels loyal to Him could live. Lucifer, whose name meant Morning Star, would from that time forward be called Satan, the wily one, and his fellow creatures called devils instead of angels. They were doomed to exist in the material universe where they could live as they pleased but could not interact with either Elohim or the spirits loyal to Him. With this act, Elohim removed from heaven all individuals who were contrary to the rules He cherished, and placed them with Satan in the material world. In so doing, He knew that Satan's power would corrupt and create chaos in the material world as it increased, but that without access to the heavenly dimension he could not use it to add more angels to his horde.

As for Satan, his first reaction was one of despair. The warm love of the Creator no longer bathed his every sense, and he felt lost without it. All was emptiness around him. For a while, his followers also mourned their loss, but as time went by, they developed a hardening of their spirits and gradually began to revel in the power they had in their own dimension to do as they pleased. They replaced love with a constant search for interesting ways to get pleasure in their universe, creating conflict and thriving from observing the consequences of it. And all this time, Satan hated his Creator and plotted revenge. He had an infinity to achieve his purpose, and he could be patient.

Satan and his fellow devils roamed their restricted space looking for a world to call their own. Driven by their need for chaos, violence, and catastrophe, they searched for one of the younger suns with their hot and fiery planets around them. Eventually they found a world, molten, hot, and desolate, orbiting two suns. They began to mold it and the other young gaseous planets adjacent to it, for their own purpose. They called it Hell. Since the only sentient beings they had contact with were their fellow defectors, they tortured and humiliated each other and devised new ways to cause havoc on the planets around them. From a distance they sensed the happiness of the angels that dwelt in Elohim's domain and their hatred and jealousy grew. They longed for the day that they could become powerful enough to overthrow Elohim and scatter His angels. With the passing of eons, they realized that this would not happen unless they had a way to get access to Elohim's domain. Satan waited for the day that he could, by guile and deception, persuade Elohim to provide him with access to His creatures. He felt that he had an advantage over Elohim because He could not lie, and he could. He would feign loyalty to Elohim and find a way to undermine His popularity among the angels.

Elohim knew of his deceit but did not challenge him on it. Satan was acting on the assumption that he was free to act according to his will. His conversations with Beelzebub had convinced him that it was possible to destroy what Elohim had created even if Elohim Himself could not be destroyed. *He will eventually want more than the insipid creatures that surround him now,* he thought. *He has created beings in His own image before which Beelzebub successfully destroyed. Time will deliver His powers to me. I will not be restricted by His ridiculous ethical rules and that will make it easier for me to triumph,* he decided.

Meanwhile, they experimented with ways to destroy small suns and other planets in the vicinity to prepare themselves for the day an inhabited planet was created by Elohim. They were unsuccessful in their attempts because they understood little of the laws governing energy and matter.

Chapter Four

The Milky Way Galaxy

4.6 gya–3.8 gya

Nine billion years passed. Elohim enjoyed the beauty and power within His creation, as clouds of dust aggregated to form dense nebulas. As millions of years flashed by, these nebulas coalesced under the influence of gravity and eventually became an assembly line for the nuclear power houses we see as stars. When we look at a dark sky at night, we see a sea of stars all of which are distant suns like our own. We only see billions of stars in our galaxy. The stars beyond our galaxy are impossible to see with the naked eye. Elohim, with His awareness of all His creation, perceived the universe in its entirety not just in the present but through the future. He saw the four hundred billion young galaxies evenly distributed through a young universe that would continue to inflate over millennia.

Dark energy, by expanding between universes, kept them apart. He also foresaw that, as millennia sped by, the trajectories of some galaxies would intersect causing them to collide and merge with each other.

The intricacies of organizing His creation fulfilled His intellectual needs and His angels filled the void that He had felt for millennia as He lay dormant as a singularity. As time went by, however, a feeling that something was lacking grew stronger within Him until it became an urge that had to be satisfied. He needed to create material living things as He had done in universes past. This time He would create the ultimate being crafted in His own spiritual image and with a material body that would be capable of existing for hundreds of years. The sun that He would provide would bathe His world with energy-giving light and heat for millennia. Elohim looked for just the right galaxy. He wanted the galaxy to have a lifespan of at least four hundred billion years before it collided with any adjacent galaxies. Although many galaxies survived such a collision without much damage to their individual solar systems, there was always the possibility that an occasional planetary system would collide with another and that the planets could be ejected into empty space. He also needed to have the perfect composition of the sun and planets to promote the formation of a world that would be habitable by living things.

He chose a medium-sized galaxy called the Milky Way to start His new solar system which would provide a planet for humans to evolve. He chose a location and caused a sun to go supernova in a nearby area in the galaxy and when it exploded it caused the overloaded area of dense dust to break up into large fragments. One of these fragments became the nebula that would form the sun. Elohim chose a chunk with about the same mass as the earth's sun. The fragment was disc shaped with a central bulge and spun on its radial axis at a rapid rate. The central area was surrounded by solid fragments of various sizes which formed a wide ring around the core. The energy produced by the spin caused the central area to heat up and become a hot liquid. The faster it spun the smaller it became, which enabled it to speed up even more. At this stage the structure was called a protoplanetary disc.

The dust produced from the exploding star settled into interstellar space and Elohim found an area that was rich in iron-60. The fine dust accompanying these elements had become incorporated homogenously into the disc. He found other elements like nickel and aluminum from other supernova remnants. Once He set the elements in place, the formation of a solar system progressed the way He had set it for all solar systems. It continued on its own, but He would be monitoring it to make sure that all necessary materials needed were provided for a planet where living beings would exist. The planet was placed in the best position for life to survive. It would become Earth, the home of man.

Over time the disc flattened out at the edges and the center formed the early stages of our sun. As the core became denser and flattened, the disc rotated faster, increasing the energy in the core. The core was composed of metals and silicates in the four inner planets. These elements were not frequent in the cloud clusters from which the planets derived. Elohim made sure He had selected a dust cluster that contained a high proportion of these substances. He wanted His inner planets to have a core that would last a long time. As the cores of the four planets liquified they became intensely hot.

The accompanying solids were now in large pieces and joined randomly with each other to form larger masses by accretion. This was an ongoing process with pieces joining the protoplanet and then being extruded from the rim. Eventually pieces accreted to form the four inner planets. When the expansion reached the frost zone, because of distance from the core the products remaining were mostly gas and frozen liquid, limiting further movement outwards. Elohim was happy to have acquired four planets with silicates and metals in the core. Because the supply of these elements was limited, the planets were all small. At this stage the core was called a protostar. This occurred about 4.55 billion years ago (Gya). Effortlessly and with exquisite logic Elohim planned every phase of the formation of His chosen planet.

Gradually, over fifty million years the material surrounding the molten core of the protoplanetary discs distributed itself from the warm inner core up to the frost line. The protoplanetary disc material coalesced into larger objects. Eventually they formed the four inner planets composed of solid materials. They were small and dense and were affected by the sun's gravitational pull. Therefore, they developed orbits that were close to the sun. Mercury, the smallest planet, was closest to the sun. Then Venus, Earth, and Mars were formed. The outer planets were composed of frozen gases, mostly helium and hydrogen as ice. They extended well beyond the frost line. Because the outer planets were rarified, they had less gravitational attraction from the sun. They were Jupiter, Saturn, Uranus, and Neptune in that order of proximity to the sun. Around 3.8 billion years ago (Gya), Elohim caused the gas planets, Jupiter and Saturn, to change their orbits. The gravitational forces that were produced caused small planetesimals to move around, thus bombarding the inner planets with material from the adjacent solar systems. Elohim directed the planetesimals containing water and organic substances like amino acids to Earth.

Elohim saw that the future of the galaxy containing the planetary system He had chosen would be safe for at least ten billion years at which point the galaxy would merge with a neighboring larger galaxy containing a trillion stars. However, because of the huge amount of empty space within galaxies and their individual suns, the chances of our planetary system colliding with another one is small. If the human species He envisioned did well and progressed in its abilities to optimize its environment and explore the world, a solution could be found for even intergalactic travel.

Elohim was content with His initial layout for the future. He could now leave things be and the physical properties He had ordained for all matters and energy would continue to shape its new solar system as they did all the billions of other solar systems in the universe. He turned His attention back to the universe at large and to His angels.

Aware of all in His universe at all moments in time, He would know when to interfere if things were not going right. Time and matter were inseparable from Him. While He focused initially on one planetary system within one galaxy in one of His universes, He had infinite potential to exercise creative skills in other galaxies and other universes that had bubbled out during the inflation of the cosmos if everything worked out with this one. They would be in other distant galaxies and the chance that the beings on Earth would ever meet them was infinitesimal.

Over the next few million years, the core of material that formed our sun became increasingly dense and energized by the various competing energy fields that Elohim had devised. Gravity, gas pressure, and magnetic and atomic energy competed to gradually produce the correct milieu for atomic fusion to be established. It took about a hundred million years for the force of gravity to cause an increasing collapse of matter within the sun's core, releasing energy as heat. Eventually the temperatures became so high that deuterium atoms began to fuse, generating thermonuclear energy which produced magnetic fields that reduced the central collapse. The nuclear reactions at the core finally settled into an atomic fusion that fused hydrogen into helium. This reaction fuels the energy radiated from the sun today. At this point the sun became a main function star.

The angels watched in awe as they understood the highly organized process introduced by Elohim. Using the behavior of gravitational mass and the electromagnetic spectrum components, He had provided an unending furnace and a source of light to Earth's inhabitants.

The sun has many layers. The molten core is surrounded by a radiative layer. Within the core, temperatures are so high that protons and electrons are released with light and heat energy. Protons are ejected from the core into the radiative layer and travel through it into the next layer. As they travel, however, they collide with each other or with electrons. This results in a trajectory of traveling back and forth within the layer. This constant change in trajectory slows them down and it takes fifty million years for them to make it into the next layer called the radiative zone.

With His perfect logic Elohim had created an energy source for His planets that would last many billions of years with little upkeep. The repeated change in course caused by frequent collisions kept the molecules from going straight into space and dissipating core energy. That would reduce the production capability to be gone in a billion years. The frequent collisions may even have created some energy by increasing momentum back and forth.

Once protons emerge from this radiative layer, they enter a rarified but still very hot region, called the convection layer. Here matter exists in blobs at the periphery, and they sink back into the underlying liquid creating a convection cycle in the hot molten liquid. This creates another delay in releasing energy into space. The temperatures are still comparable to those in the radiative layer with varying temperatures and energy content.

The outer surface of this convective layer forms a thin skin at whose outer interface light and heat energy escape out into the sun's atmosphere making it visible to us from Earth. Because of this phenomenon, we call this surface layer the photosphere. At this interface all forms of energy forces contained in the electromagnetic spectrum also escape into the sun's atmosphere. From there they move outwards toward the planets of our solar system and then into interstellar space. The temperatures just beyond the photosphere are much higher than those in the convective layer. This marked increase in temperature is not fully understood as we struggle to learn more about the intricate mathematics and physics Elohim built into the formation and maintenance of His suns. We can only be reassured that He has provided us a distant future and that humanity will prove worthy of His care of us.

By sixty million years, the planets settled into almost circular orbits around the sun. Mercury, Venus, Earth, and Mars were exposed to the intensely hot radiation from the sun. On Mars, all substances that could be vaporized to gas were lost in surrounding interstellar space. It could not hold an atmosphere on its mostly metallic surface. As our solar system evolved, a ring of solid and gaseous debris accumulated and orbited the sun beyond the orbit of Mars. This persists today as the asteroid belt, an accumulation of frozen chunks of matter trapped in orbit by the gravitational pull of the sun.

Beyond the four inner planets, Jupiter, Saturn, Neptune, and Uranus also rotate in circular orbits. The force of gravity kept the orbits much closer to each other then than they are today. Beyond the orbit of Uranus there are several concentric belts of frozen solid objects, some of which are the size of small planets. The layers decrease in density and particle size as distance from the sun increases. Elohim engineered changes in the synchronized movements of the planets and their accompanying rings of orbiting frozen asteroids.

Over the next five hundred million years, Elohim caused a shift in the orbits of Jupiter and Saturn so that they developed a two-to-one lockstep synchrony with each other. Until recently it had always been assumed that in the beginning Jupiter displayed three orbits for every two of Saturn's. Supercomputers however have demonstrated by enabling more accurate measurements that they have been in a two-to-one resonance right from the beginning. Three to one would have been a chaotic history when it is compared to other pairs in other constellations. The effect of this synchrony was an outward gravitational force that pushed on Uranus and Neptune. Neptune, with a bigger mass than Uranus, migrated outward faster until its orbit brought it in contact with the fringes of the solar system. Here, it collided with frozen planetesimals of diverse sizes, sending them flying at a high velocity both outward into interstellar space and inward toward the inner planets. Many collided with Uranus and the fragments produced bounced inward, reaching the regions of the inner planets where the new earth resided. Many of these objects wound up as moons around the gas giants. Others were hurled outward by the massive gravitational pull of Jupiter to join the planetesimals in the Kuiper belt. Smaller particles settled into more distal orbits in the fringes of this satellite system to form the Oort cloud.

Over millions of years Uranus and Neptune adjusted their orbits until they were roughly circular again. The miraculous result was a single plane of orbital movement for all the planets and the asteroid belts that accompanied them. They were positioned at an angle of thirty degrees to the sun. This is very similar to the Ecliptic line. This is an orbital plane in which most eclipses occur. Chance or “Mother Nature” did not do this. Elohim with foresight knew that these changes were essential for the establishment of multiple processes that would eventually shape the trajectory of His evolutionary process for life forms.

The violent displacement of the planetesimals orbiting at the fringes of the solar system by Neptune caused a massive bombardment of the inner planets. This resulted in cratering of the planets. When the fragments were large enough, they broke off portions of the planet and these fragments orbited their parent planets as moons. One such fragment carved out our moon. Elohim made sure that earth received many fragments containing the ingredients that would form water vapor. As the planets formed, solar winds containing charged particles and energy swept the remaining matter between the orbits of the planets out into interstellar space leaving the solar system with dark unlit space around the various bodies orbiting the sun.

By 3.65 billion years BCE (Before the Common Era), the inner planets Mercury and Venus had become uninhabitable hot fiery worlds. Mercury, as it was closest to the sun, lost its atmosphere as it evaporated under the fierce glare of the sun. There was no chance for water vapor or rain to accumulate onto its solid metallic surface. We know though that in the depths of a crater at the poles, where the radiant heat of the sun does not reach, there is water ice today. Venus has a mass and size comparable to Earth. Because of its proximity to the sun and its very thick atmosphere, which causes a profound greenhouse effect, it is the hottest planet in our solar system. Its atmosphere consists mostly of carbon dioxide with insignificant amounts of nitrogen and a small amount of water. Beneath the thick atmosphere it probably has a liquid core with a thin crust. Both Venus and Mercury have persistent volcanic activity as their molten cores vent magma to the surface.

Elohim, having selected our earth as its chosen planet for life, made sure that its size, substance, and position in terms of distance from the sun would produce a planet that would be an optimal milieu for life. He also made sure that the little planetary disc received all the organic chemicals needed to eventually form the ingredients of carbon-based life forms. Such organic molecules were already plentiful in the cosmos and made up the asteroids and comets in the Milky Way, as it came to be called, and in other galaxies and nebulas.

As solid elements containing organic molecules entered the planetary disc, Elohim made sure that they were retained in the collection of solid fragments that had already begun to aggregate to form the earth. Amino acids were an important part of this material. He also selectively engineered the aggregates that were rich in nickel and iron to occupy the core of the planet. It would be a repository for storing heat that comes from the sun. When colder climatic conditions occur, the core radiates heat outward to maintain warm temperatures at the surface of the planet. The uniqueness of Earth as a perfect repository for life indicates that this is not a chance event. Today, using our telescopes in the sky and on the surface of the earth we detect and analyze the constituents of the many new planets that we are now able to detect. To date, not one planet other than Earth has been found to have all the qualities that would support life, and none has been shown to contain large oceans.

Beyond the asteroid belt that circles the inner planets lie the four gas giants. They are composed of mostly helium and hydrogen, the lightest elements. They were swept outward by the early solar winds and eventually gathered into masses of frozen gas with their warmest temperature at the surface ranging from -160 to -360 degrees Fahrenheit. Nearest to Earth is Jupiter, which lies approximately five times the distance that Earth is from the sun. It has multiple moons. Almost no water vapor exists in its atmosphere. It continues to march in resonance with its partner Saturn which is next at 9.5 AU (astronomical units) from the sun. Saturn is remarkable for its large rings consisting of frozen particles that are visible from Earth. It has over thirty moons, the most remarkable of which is Titan. Titan is the only moon in the solar system that has an atmosphere and that has liquid water on its surface. Elohim saw that as the solar system aged and the sun became hotter, Titan could potentially serve as a planet where life could evolve anew or be transplanted from the earth.

Elohim rested after He had carefully engineered the proper milieu for life to develop on His little blue planet. For two billion years, Earth remained a fiery molten ball with a liquid surface. Around 4.4 billion years ago, a solid crust began to form. We know this from the examination of zirconium crystals in rocks of today. Prior to that we assume that the earth had a fluid crust that moved and shifted. All geologic and fossil formations therefore began about 4.4 billion years ago. The period before this is called the Hadean and it was a hot fiery time of about two hundred million years. During this time Elohim laid the foundation for the proper formation of a hot fluid core surrounded by a mantle of metallic material that circulated outwards from the core taking heat to the surface. With the zirconium and iron He maneuvered into the core, when it was just a planetesimal, there was plenty of raw material to fuel a hot furnace that would last billions of years.

It was a time of heavy bombardment by asteroids and fragments from neighboring planets. It provided Elohim with lots of opportunities to seed the required materials for life into the fabric of the earth. Toward the end of the Hadean Eon, four billion years ago, He began to lay the groundwork for life to evolve. First, He induced water, which He would make an essential ingredient for all life to exist, to be retained on the third planet in large quantities. The earth's crust was beginning to form and there was much out-gassing from the mantle below it which contained water. He made sure it stayed under the plane of the mantle and did not fizzle out into space. He also ensured during the bombardment of the solar system that occurred when

Uranus and Neptune traded orbits, that the earth received more than its share of frozen water. Further additions were engineered via collision with extra solar asteroids largely composed of water.

The mantle already contained carbon-containing molecules, and these were mainly carbon dioxide and carbon monoxide, both contributing bound oxygen to the atmosphere. As the atmosphere accumulated carbon dioxide and water vapor, a greenhouse effect was induced. The early sun was weak in comparison with our sun today. By four billion years ago, the hydrologic cycle of condensation, vaporization, and rain was initiated. The delicacy of this process is still not fully understood, and many intricate scientific theories exist as to how this process started and then led to the formation of a surface composed mainly of oceans. A similar process of trapping gases in the atmosphere of Venus resulted in such a thick atmosphere that the surface became unbearably hot and inhospitable to the evolution of life. The accumulation of greenhouse gases at just the right amount initiated cooling at the earth's surface leading to condensation of water and the formation of the early oceans. You will find this kind of serendipity pops up quite frequently in reference to happenings on Earth, compared to other planets.

Beelzebub watched Elohim as He went ahead with His creative process. Elohim had experienced several such universes come to be and then fall apart because of Beelzebub's interference. By devious methods it had destroyed all that Elohim had constructed and created chaos. In more recent universes, however, he could not take any responsibility for the destructions. It had been the work of Elohim's creations. Beelzebub had to admit that at least one time it had been so disturbed by the heinous deeds of the beings that Elohim had unknowingly created, that it had secretly joined forces with Him to destroy them. It obtained great pleasure doing it and was surprised when Elohim still mourned their loss.

Beelzebub snorted with derision. What chance had love, compassion, and justice, which Elohim championed against its own offerings of hate, jealousy, greed, and power, to win over the frail, easily deceived material constructs that Elohim would cause to evolve. All Beelzebub had to do was devise a way to have access to the creatures. Imprisoned in an unknown area, it had no way to access the planet and wreaked havoc on it. It doubted that Satan had as much experience with creating chaos as it did.

Elohim, despite His contentment with what He had done so far did have some concerns. He remembered sadly that His hidden alter ego, Beelzebub, was temporarily subjugated by the power of His love, but not eliminated. He remembered the beings He had created in His own image in previous universes, programmed to love and be loved. Each had succumbed to the call of the small amount of chaos that had unavoidably been included in their fabric. Elohim was willing to face that challenge again, but first He would create a suitable environment for His creatures, one that enthralled them and kept them searching for answers as they evolved. For along with the ability to love and create, they would be endowed with their Creator's endless search for new enigmas.

Chapter Five

The Beginning of Life:

DNA and RNA

3.8 gya–2.5 gya

Six hundred million years have passed since the beginning of our solar system. Approximately four billion years ago, planet Earth consisted of a persistently hot and active crust that was covered by large areas of shallow oceans. Beneath the crust, magma flowed in an endless convection cycle from a hot, solid iron-nickel core to the semi amino acids, forming the building blocks for all forms of life. Elohim examined all the amino acids available to Him in the milieu of the early earth. He had a choice of more than twenty. But, to simplify and instill order into the genetic process, He decided to select only twenty of the available ones. He also saw that there was a dichotomy of structures in the existing amino acids. All individual amino acids existed in two mirror images of each other. He also decided that the various macromolecules that form the structure of living tissue, such as nucleic acids, proteins, carbohydrates, and fats, would have only one kind of structure—left-handed or right-handed molecules, but not both. This phenomenon, which we call homochirality, is seen in all biologic macromolecules and seems to have been a prerequisite for life as we know it to evolve.

It is of interest that synthetic amino acids created in a test tube today are always a racemic mixture of the two chiral forms, and we assume from this that racemic mixtures were the milieu of the pre-life world before Elohim changed things by introducing homochirality in amino acids. This poses problems when scientists try to produce life from scratch in a test tube. All current organic chemicals are found in two forms that are isomeric images of each other. That means that they reflect light either to the right or the left but not both. Right off the start it becomes impossible to duplicate the process used by Elohim. They would have to first produce amino acids that were homochiral. The mono-chiral type, created and used by Elohim, do not exist in our world and are difficult, if not impossible, to produce in pure levo- or dextro- forms.

Unable to mimic Elohim's process, scientists first focused on looking at reasons why there were homochiral populations of amino acids and other chemicals in the earth's environment at that time. They claim that the geologic conditions they found indicate dramatic electromagnetic changes in the atmosphere that could have affected the form of organic molecules making groups of chemicals to be mono-chiral. They hypothesize further that the conditions could have been favorable enough to spontaneously allow amino acids to *polymerize* into nucleic acids and then proteins. Furthermore, scientists make these claims not with clearcut results from experiments or from supportive geologic data.

As the next step, an experiment by Thiermann and Jarzac is cited. They produced synthesized amino acids in a magnetic field and found that by varying the magnetic field they could produce amino acids high in either dextro- or levo- forms, but not pure collections of one or the other. Many hypothetical explanations are put forth on the basis of this temporary flux in electro-magnetic states.

Forterre, Filée, and Myllykallio focused on the evolution of DNA and DNA replication machineries as the preliminary process for creating life. They produced a very complex hypothetical discourse on the many possible connections to the evolution of life (DNA). It focused on their opinion that the evolution of DNA was linked to the presence of viruses on the early earth. In conclusion, they recommended more research on the role of viruses being the source of DNA. Also, they recommended the involvement of scientists working in comparative genomics to look at possible theories from their perspective.

Let us get back to Elohim two billion years ago. As He had done for physics, Elohim laid down the rules for organic chemistry and inorganic chemical processes for all His creation. His goal was to establish a world in which all energy forces could transform from one to the other with little loss of energy. When the time came to start the existence of living things, everything would be ready.

His plan was to introduce a prototype prokaryotic cell as soon as He felt the time was right. He was knowledgeable about all aspects of cellular function. In His many universes the organization of first life and the ensuing rapid stepwise increase in sophistication had been very similar. He would continue to use His ability to produce a world for His new sentient beings that would keep them looking for answers. It would be a planned creation with everything ordered and logical. He realized, however, that with the forces of Satan and Beelzebub eagerly waiting in the background to wreck His plans, there would be ups and downs.

As life evolved, He continued to add regulatory instructions to enable continued dynamic intracellular processes. He chose to start His creation at a time when the earth was still chemically active. All ingredients that were needed were easily available. He made sure that biochemical components were evenly distributed as they would soon be needed.

Another five hundred million years passed in a flash. Elohim watched over the developing earth and felt conditions were right for the next step. He was ready to create the first living cell and planned to devise logical and efficient energy systems to provide energy to fuel chemical reactions. Into this ordered chemistry He would introduce a mechanism for the evolution of a variety of distinct kingdoms of creatures.

With the cooling of the earth and its oceans, many shallow ocean beds were composed of crystalline clays that underwent consecutive cycles of wetting and drying. Complex organic molecules containing carbon, hydrogen, oxygen, and nitrogen were already present in the earth's solar system and in many areas of the early earth. They had been formed early in the universe from the interactions of stellar gas and dust with cosmic rays. During the heavy bombardment of the early earth by interstellar and intra-solar objects, Elohim made sure that the earth received more than its share of these molecules. They would provide the carbon required for carbon-based life. Also present in the substance of the earth was a full complement of amino acids formed by the interaction of various physical forces, like lightning, on organic molecules.

Elohim had already ensured availability of amino acids displaying homochirality and He now selected the twenty He needed to provide the full spectrum of living things. Today, we see as the end result, the process of protein synthesis. The transformation of two vital ingredients—amino acids and nucleotides—combine to form the nucleic acids RNA and DNA. Every amino acid is composed of a carbon-containing acid group and a nitrogen-containing amino group. Where the existing twenty amino acids differ from each other is in a special group called the R group.

Each amino acid has its own unique R group. This property makes amino acids a good fit for a biological code for genetic variability. Amino acids in chains are called peptides and they are the foundation units for proteins. Proteins can be small with one or two amino acid chains, or they can be enormous clusters of amino acid chains called polypeptides. They are composed of different amino acids linked by weak hydrogen bonds. Depending on the nature of these bonds, thousands of different molecules called proteins exist in the cells of many tissues where they are used for cell repair or as enzymes. They are manufactured according to need and used immediately since they tend be unstable.

Since much of the process is repetitive and automated, there are few mistakes, but in case there were problems, Elohim introduced into each cell instructions for what to do if an error occurs. For example, if during transcription of a protein an error occurred, there are instructions for the cellular mechanism to correct the error by cutting off the segment of the messenger RNA that was corrupted. This is followed by replacement with a normal segment.

Elohim thought about the molecules called RNA (ribonucleic acid) and DNA (deoxyribonucleic acid) that He had introduced. They were very similar in structure. This similarity would enable them to be used as templates to make copies of each other. DNA is a helical double-stranded molecule. Ribonucleic acid (RNA) is always single stranded. If you took apart the two strands of DNA, one of them would be identical to an RNA strand. The other would be a good template for creating an RNA molecule. The nucleotides in one strand would be complementary to the other. There were two other differences in the strands between DNA and RNA. Deoxyribonucleic acid (DNA) has thymine as one of its bases instead of uracil. Ribonucleic acid has a substitution of uracil instead of thymine. Also, DNA has deoxyribose as its sugar component while RNA has ribose.

Elohim decided to create a process where a cell could initiate the process of protein synthesis using one strand of a DNA molecule as a template for making RNA. He chose to set up the process in the first cells He created. He documented the process in the cell in a retrievable form. During cell division it is passed on to the new cells. He created this first cell with copies simultaneously produced for multiple areas of the world. The information outlined the responsibilities of each component in the cell. It would be passed on like all genetic material to every daughter cell produced. The DNA would be in the form of a chromosome. When the DNA molecule is used as a template to produce a pre-RNA molecule it is first loaded with bases by RNA polymerase, an enzyme. It only needs the addition of appropriate amino acids, using a reading frame to determine which amino acid attaches to which base.

This is a process that is called translation. It includes a complex procedure that vets the molecule to make sure it is acceptable. This is performed in the ribosome. The process of protein synthesis is one of the most important of Elohim's creative expressions. It is relatively simple in prokaryotes (unicellular cells) but becomes complex in eukaryotes which would be the next advance in life. Replication becomes much more complicated in eukaryotic cells.

Having reviewed all the steps He was going to use, and making sure all ingredients were present, Elohim started the creation of the first prokaryotic cell. Once the process was complete, He transported the cells to various locations which would provide a suitable environment for life to begin and continue to develop. He prepped the cells by adding needed contents into them. He made a mental list of what He would need to jumpstart a live prokaryotic cell. It would be a while before the earth was ready for eukaryotes.

He was pleased with His creation of first life in this manner. He hoped the act would be followed by a rapid spread of new living creatures into a world which was pleasant to live in. The Precambrian Era, approximately six hundred million years ago (mya), had a warm, pleasant climate optimal for the beginning of life. Into this world Elohim planned to introduce the basic living cell which He enabled with the knowledge of RNA and DNA synthesis. It would start with the process of producing a molecular copy of the two DNA strands. One of these strands is later used as a template to make proteins. When a protein is needed, the molecular DNA copy will localize the area of the gene that is to be replicated. The entire DNA molecular copy is stored in the ribosome present in the cell and is available to be used to set up the production of a pre-RNA molecule.

This molecular copy is used to create a template called messenger pre-RNA, which is composed of specific bases, phosphate groups, and ribose groups. The addition of appropriate amino acids completes the process. So, protein synthesis involves two stages. First a protein precursor, called messenger pre-RNA, is produced by adding nitrogenous bases to ribose molecules in the base- pairing order arranged in the template. Nitrogenous bases have an inherent preference for a partner base. The messenger RNA itself is not yet a protein. It is a pre-RNA molecule which will only become a protein when an amino acid is added to each nucleotide that is present in the messenger RNA template. This is done by RNA polymerase, an enzyme which loads the bases with the correct amino acid for each base. The information for deciding which amino acid fits with which base is provided in a table called the reading frame. It is essentially a coding box. It enables the selection of an amino acid that is compatible with the triplets of nucleotides called codons. Once the RNA polymerase has completed this process called translation, the new molecule will need to be vetted before it is released. In prokaryotes this is a mere formality and begins as an ongoing coordinated process with amino acid loading. One or more polymerases will deal with translation as the RNA pol is loading the amino acids. The process completes and the new protein is picked up for immediate use as an enzyme or a protein that is needed for constructing parts of the cell.

The same process in eukaryotes is much more complicated. Two of the extra bases become a *START* button and three bases act as *STOP* buttons. One base becomes the AUG (adenine, uracil, guanine) codon which is involved with finding the methionine-controlled start location. The reading frame for use in translation is set by the AUG codon near the five-prime end of the mRNA. Sometimes there may be more than one amino acid to a base. The coding box, called a reading frame, is used to find out which amino acid goes with which nucleotide triple codon.

During both RNA replication and DNA replication there is a structure called the *template*. DNA replication and RNA replication is responsible for producing and maintaining the template. The template will be mentioned often throughout the process of protein synthesis. It needs an explanation. When reference is made to the *wall* or *backbone*, they refer to its wall-like appearance with upper and lower edges, which is what each strand of a DNA molecule looks like when it is unwound from its helical shape. The upper and lower edge are actually two lines of triplet codon heads, and the template is the space that lies between them. Each border is formed by the objects within it. For the left edge this comprises a phosphate molecule which sits at the extreme edge. Then placed a little to the right is a pentad molecule of ribose sugar. This position is critical because it allows the placement of the correct codon close to carbon-3 in the ribose. This is where the phosphodiester point is activated to produce energy for the chemical process by releasing a phosphate group from ATP (adenine triphosphate). It results in the formation of ADP (adenine diphosphate). By this time the codon is already in position floating within the template just to the right of the ribose. This process is duplicated on the opposite edge resulting in the space between being formed by the two rows of codon tails almost touching. The sugar has to be ribose for RNA or deoxyribose for DNA. The energy being produced in the area of the template makes it appear tangible. Template features are similar to the template produced to form messenger RNA and for replication of a DNA strand. The two edges of this template are called backbones because the nucleotides attach to them, hanging free between them.

The nucleotides are of two types. There are three pyrimidines that each have one pentagonal ring, and two purines that have two rings. The purines' two rings are attached along one side and are composed of a pentagonal ring and a six-sided ring. Adenine is partnered with thymine and they form two hydrogen bonds; cytosine and guanine form three hydrogen bonds. These attachments with the chemical energy of the bonds keep the rectangular template open and movable. Each row across consists of one purine and one pyrimidine.

Another perspective would be that on either side of the rectangle of triplet codons runs a chain of a phosphate group linked to the pentagon of a ribose or deoxyribose sugar. They run downward from adenine at the five-foot end on the left and upward from the five-foot end on the right. The phosphate-sugar chain forms the backbone on each side. The only difference between a DNA template and an RNA template is that DNA has thymine instead of uracil and RNA has uracil instead of thymine, and the sugar is deoxyribose for DNA and ribose for RNA.

For DNA the nitrogenous bases are adenine (A), guanine (G), cytosine (C), and thymine (T). Each is named according to the base it carries, and each is inserted into the backbone described earlier. But a problem arises. The bases are fussy and will only connect to one partner. Adenine will only connect to guanine and thymine will only connect to cytosine: A-G and T-C. This is a step toward forming a code that can produce a huge number of proteins just by varying the pattern of the nucleotides. The bases need to link to the backbone between a phosphate group and sugar. This is an energy-carrying link called a phosphodiester bond. A nucleotide attaches to one edge of the backbone. The other edge of the template will carry its own set of nucleotides attached in the same way. They are spaced accordingly. The big difference is that the left-sided backbone moves from top to bottom and the right-sided backbone moves from bottom to top. As long as the sequence of nucleotides is started correctly along its backbones, the nucleotides will be a match. This is because each half of the template lengthwise is complementary to the other. Each nucleotide will always find its favorite opposite to it. In addition to this unusual movement, the template is involved in creating the energy exchange that occurs at the phosphate, deoxyribose, and nucleotide junctions. This provides energy for the process of protein synthesis.

This process sounds complicated to organize but it is not. Remember, the right backbone moves from the bottom to the top. The left moves from top to bottom. They are two strands of DNA that normally exist as anti-parallel double helices. This means that when the two strands are opened up and stretched, they move anti-parallel to each other in opposite directions. This makes it possible to synchronize their movements. The energy-producing joints are formed by connections between the three structures. The numbers refer to the carbon atoms which are part of each junction. They are located on the pentagon that forms the body of the sugar molecule and go from one to five. It has five carbon groups that attach to OH groups that attach to the phosphate and they go from three to five. The strands are anti-parallel and therefore the right one goes from bottom to top in a three-to-five prime direction and the left starts the nucleotides in a five-to-three prime direction. The result will be two complementary strands. The right strand is called the leading strand and the left one is called the lagging strand. The leading strand is the DNA polymerase that starts because it is programmed to always go from three to five. This ingenious setup on the part of Elohim made the matching of nucleotides smooth and error-free.

Elohim starts by constructing the cell prototype. He creates a small bubble-like container that has a wall that is stronger than it looks. He has strengthened the cell walls with peptidoglycans so that it keeps its shape. Then He starts adding the structures to His list that are needed for the initial step. First, He designs and adds a structure called a ribosome. Ribosomes are structures within the cytoplasm in both prokaryotes and eukaryotes that perform some of the tasks of protein translation. Elohim chose to introduce all the parts of more advanced cells would have been in their nucleoid and cytoplasm, directly into the cytoplasm of the cell. This would be a model used in prokaryotic cells (single cell organism related) but contain some eukaryote features.

Elohim started by creating a molecular copy of the DNA molecule which would provide the template for producing pre-mRNA. The enzyme RNA pol adds the correct nucleotide for each position. This is repeated until the template is filled. This is the first step in creating the protein product. Messenger RNA is a template for producing a variety of templates that will become proteins after the addition of amino acids. The RNA pol checks to see if a protein copy had been requested. If it has, it goes ahead with loading the amino acids. It does this by starting its insertion at the first position at the start codon. A reading frame is posted at the site. *RNA pol* refers to the ability to choose every amino acid to be added. Once it completes the entire segment, the product is ready to be delivered. At this stage, it is a messenger RNA molecule representing a specific protein.

The molecular copy that preceded the addition of amino acids, called pre-mRNA, is stored in the ribosome for future use as a template. It could be one peptide or multiple connected peptides. Once the actual protein is configured, it needs to be used quickly, or it will disintegrate. Each gene is a specific coded segment. They are repaired by cell occupants and can be achieved by removing the affected segment and adding a new corrected template with the proper nucleotides and the correct amino acids.

This is the first time the process is performed anywhere. The DNA in this first performance is a passive participant in the process. It is not yet in charge of the cell as it will be later. The orchestrator of the whole process is the promoter. It starts and stops the process by manipulating start and stop signals, though it does not play a major role in the actual synthetic activities, as it acts as a coordinator. The next important participant is the enzyme RNA polymerase. It will do all the work of adding nucleotides to the RNA template. The template is defined by the promoter who creates about ten parts from the RNA segments that are being transcribed will start. It will manage the stop and starts. One of the strands of a DNA molecule will be laid out with the backbone accessible. It is called the leading strand. The second DNA strand will not be participating. It is called the non-leading strand. A segment of mRNA will move along the DNA segment in a five-to-three prime direction, breaking up and then closing the template as it attaches bases and moves along. In eukaryotes this is a much more complicated process. To initiate the process, tRNA polymerases are multiple and need special transcription factors and their own promoter to accomplish their activities. The holoenzyme will direct the rest. It also takes care of reconnecting the bonds so that the loop would close behind it.

Next, He added a single ring of double-stranded DNA that He created to serve as the template. Then some nucleotide molecules and amino acids were created. All are needed to build the protein this first time. He kept ready a molecule of an enzyme called DNA polymerase and had another called RNA polymerase available. Each of these is a package of isoenzymes that handle various stages of the protein production cycle, from initiation to ending. The DNA polymerase will be used to create a molecular copy of the DNA molecule. It will be required to provide a template to synthesize proteins or to provide a molecular copy of the molecule for storage in the ribosome. The RNA polymerase would then be used to create RNA proteins.

Once this happened, DNA polymerase I, II, and III became active. The two strands each consisted of a backbone of alternating phosphate groups and ribose groups with nucleotides projecting down from the backbone. The connection to the backbone needed an OH group which would connect the ribose to the phosphate group by forming a phosphodiesterase bond with the nucleotide.

Let us get back to Elohim's first production of life. The first thing that happened was the addition of an enzyme called helicase. As soon as this happened, a segment at the base of the fork became the promoter. The DNA polymerase wrapped around it and followed the directions of the promoter which was a special small strand of DNA.

The promoter chose a segment of the circular chromosome to be the starting point of replication and was about two hundred fifty base pairs long. It also marked a start button for adding the first nucleotide. Next, it added an enzyme called helicase which began to break apart the bonds that kept the helix going, starting in a specific initial replication activity which began with the unwinding of His double helix into two longitudinal segments of DNA, a leading strand and a lagging strand. The two DNA strands started to elongate, separating into two strands and forming a structure that was Y-shaped, like two prongs of a fork.

He kept the amino acids ready for use. They are all needed to build the protein this first time. He kept ready a molecule of an enzyme called DNA polymerase and had another called RNA polymerase. All these agents will be needed to finish this amazing transition. They are essential for the process to happen, especially the first time. It is much simpler in prokaryotes. This process alone is one of the miracles that people say does not exist. To quote Albert Einstein, "The fact that the cosmos is incomprehensible, that it follows laws, is worthy of awe. This is the defining quality of a God who reveals himself in the harmony of all that exists." Watching the process happening we can experience this miracle.

Once He added all the ingredients, Elohim decided that His first step would be to build a molecular copy of DNA. Once created, it would be like a photocopy of an image that can be stored and used to identify individual genes. A molecule of DNA is a complex chain of genes which can be in the form of a peptide or a chain of peptides. Molecular templates can be stored in the ribosome and used to make protein.

He initiated this step once because it was needed for all RNA replication activities. Protein templates are made by adding amino acids and other chemicals to the messenger pre-RNA. The procedure produces a protein copy of the genetic segment the cell requested. Once it is ready, it will be taken to the cellular area that needs it by transport structures. In His first cell though, only the essential segments were made. Once it was clear that they had established themselves, Elohim started working on additional features that would make the process easier and suggest maintenance actions that would reduce errors.

Life should never have started. The raw materials were not there and if Elohim was not the one who established the process for the first time, then there is no one to claim credit. Remember, that until the first appearance of DNA on land, there was no life at all, not even viruses. In fact, Elohim or someone like Him was all alone until the first mammals came to be. For most of the earth's life, sea creatures were the only inhabitants.

The two strands were not identical in the mechanism of loading nucleotides. One strand is called the template strand

and is the one enzyme that DNA polymerase III began applying nucleotides to. It started doing so and smoothly added the matching nucleotides. The nucleotides had to be DNA nucleotides with thymine instead of uracil, and deoxyribose instead of ribose in the backbone as the sugar. This is because we are making a DNA molecule. A primer was added to the base of the fork to the free nucleotides to prevent the ends from rejoining. The enzyme proceeded in a five-to-three direction related to the clockwise or anti-clockwise movement of each strand. This would be a natural unwind movement to the spiral.

The template of ribose and phosphate groups was needed to have a spot where they both connected via an OH group to form a phosphodiester bond when the proper nucleotide was added. This would be the job of DNA polymerase III. An OH group was present on the template strand which was going in a five-to-three direction and DNA pol started by placing the first nucleotide at the OH group as directed by the promoter. Because molecules are being built in a five-to-three direction, the polymerase can only form in this direction. During this process RNA polymerase is firmly attached to the strand by a sliding clamp.

The design did not go smoothly with the lagging strand and two problems occurred. The lagging strand did not have an OH group at the start zone. This was because the two strands have an anti-parallel relationship to each other. The enzyme primase jumped in and introduced a primer strand which was five to ten bases long. Primase created an OH end and connected the main strand to the phosphate end via a phosphodiesterase bond and the end of the next nucleotide. Once this initial connection was made, the nucleotide addition could continue in sequence but was still slower because it was going in a three-to-five direction, which is away from the origin of the Y. This caused problems at the base of the fork.

The leading strand was winding faster than the lagging strand, and this put tension in the leading strand. At this point an enzyme called topoisomerase stepped in and made small cuts in the stretched spiral strand and balanced the strands, then joined together the cut ends. The lagging strand now continued adding nucleotides in small batches and then anchored the segment using the clamp that polymerase III had used.

Both strands were now smoothly adding nucleotides. There are four nucleotide bases available which are used in both RNA production and DNA production. Some information on nucleotide bases used may be helpful. A nucleotide is a string of nucleoside bases and both single strands of DNA can act as templates. One would be used as a template to make RNA. If a whole new strand of DNA was being manufactured, then both strands would participate in the sequence as described in the procedure Elohim used. The result would produce two DNA molecules each with two strands. In each DNA molecule, one strand would be from the original DNA molecule while the other would be from the new DNA molecule. Since we are making a copy of a DNA molecule, the nucleotides need thymine which is substituted for uracil. So, the nucleotide bases for DNA are thymine (T), adenine (A), guanine (G), and cytosine (C).

Elohim had His genetic code for making the proteins that were essential for life to begin and continue timelessly. From that point on it became established and is accepted as is today. The genetic code is a combination of nucleotides (A), (C), (T), (G), and (U), and they form codons of three to connect each to an amino acid that is compatible. Sixty-four possible combinations of nucleotide triplets are available to match with twenty amino acids. This leaves more triplet codons than we need. Some of the excesses are used for special functions like start and stop.

It is called the genetic code because it reveals that DNA produces RNA and RNA produces proteins. Each protein is named according to an alphabetic formula including the particular nucleotide and amino acid that it is formed from and this has not changed since it began two billion years ago. The double-stranded DNA molecule can easily be replicated from a strand of RNA. In turn, the messenger RNA produced can be converted to proteins.

The chemical steps to accomplish this process are complex. As it is described, people may think that someone in the lab is maneuvering the process. In reality, these processes of protein production have been functioning since the first cell was launched by Elohim. When life advanced to eukaryotes He added the extra steps and enzymes needed to create the first eukaryotic cell. And He added new instructions that the enzymes should follow. Soon after a new cell is born it can fabricate proteins which are used for cell growth, cell replacement, and making enzymes, which are required to increase all the complicated metabolic processes that are run by the cell. This routine has not changed at its core, for billions of years, following a routine that has evolved in the cells. There is a complete system of quality control and numerous enzymes to guide the process.

An enzyme is itself a protein. I assume that Elohim created it because there is no one else that we can credit with it. Chance and nature you might say, but the process developed too rapidly for natural selection to be responsible. It looks like it may be problem-prone, but it has few problems and, if any occur, the DNA and the enzymes in the cell know what to do. Who is *nature*? And why has nature not introduced new and effective solutions to problems we have with a whole variety of problems related to it today?

The DNA in the cell manages the entire process including informing the cell what protein is needed and where it is needed. In addition, if an error occurs and one of branches of the protein is structurally wrong, the enzymes in the cell can remove the erroneous part and substitute a correct chain. Even more astonishing is the fact that the genetic code described has existed since the first cell was created and is used in almost all living creatures. So, it is the basis of life and it came with a manual of instructions.

Multiple enzymes placed in the cell by Elohim, established simple metabolic cycles within the cell because without an energy source chemical reaction would not occur. The entire process has survived with little change since about two billion years ago when the first cell was created. Of course, Elohim needed to make changes as more complex body parts developed. As specialized organs developed, they would have specialized cells that performed specialized functions. For example, a liver has unique cells because it is involved in many metabolic functions. Replacement liver cells, however, would be manufactured in the liver and supervised by its DNA. Without the initial DNA and RNA this complex process would have gone nowhere. Fortunately, Elohim knew exactly what was needed and had the ability to create any reagents He did not find in the environment.

I should let you know that we do not fully understand how these homeostatic processes that prevent chaos in biologic systems are accomplished. What introduced the problem-solving measures like the cutting off of corrupted segments of a newly manufactured protein? It is not just the processes themselves that pose a problem. They are mostly well-understood chemical reactions. What is amazing is how independent cells are in taking care of themselves and their responsibilities. It is not consciously done. The processes are automatic and backed up by alternate actions if the first level of intervention fails. The process of protein synthesis is a good example. Scientists have gradually over the years studied cells using electron microscopes and watched the process happen. They are familiar with the steps in protein synthesis that have been outlined. The process was studied by scientists, but they did not create the process or modify it in any way. They visualized the process and described it.

Another amazing phenomenon about the genetic code is that it is the same in all living things, so it is very ancient. It looks like something humans would devise but they did not. The code was discovered and analyzed, and it was decided by someone or something to have triplet codons of nucleotides. Doubles would not provide enough codons to match all the amino acids. With triplets we have more than one match per amino acid which means two codons could be used. Because of this, the code was felt to be degenerate. However, it did not seem to matter! An extra protein might be created but it achieved its purpose. Elohim had done His magic. Today, the genetic code is used by all living things. I doubt that it is a product of nature.

Let us get back to the DNA molecule. When the DNA polymerase is done with the template strand it takes over from the primase and restores the remaining section of the lagging strand with DNA nucleotides. Immediately, DNA polymerase I, which is an exonuclease, starts the repair on the RNA primed portion of the lagging strand. It breaks down the RNA portion of the strand and replaces it with DNA nucleotides. Any nicks that are left empty between the replaced strands are filled in by DNA ligase which also catalyzes the last phosphodiesterase bond in the lagging strand. It fills gaps in the backbone of ribose and phosphate groups where the Okazaki fragments were connected so that there are two smooth strands with all the correct nucleotides.

When both strands are completed, the promoter checks the strands to make sure there are no mistakes and if there are none it requests a release of the DNA molecules. At the separation signal the two new strands of DNA, each with a backbone on each side, are now two DNA molecules. Now these two DNA molecules each possess a new and an old strand that are complementary to each other. This is called a conserved synthesis. The molecule which is in protein form will be converted to a molecular copy. It will be stored in the ribosome to use in DNA replication or protein synthesis. During this process, ATP (adenosine triphosphate), a nucleotide, which is part of the phosphodiester bond in the backbone, is engaged, leading to ATP changing to ADP (adenosine diphosphate) with energy release. Elohim, always vigilant, had taken care of the energy needs of the chemical processes.

He began to guide the RNA polymerase to produce a messenger RNA molecule. Elohim had empowered this molecule and others like it that He would introduce later to do amazing things. His natural tendency to build things that took care of themselves was apparent even before He started. He added essential chemical needs, but He also started the process in a systematic ordered way. Elohim viewed the portions of the DNA molecule that would be the first templates of ribosomal RNA and messenger RNA. He selected portions of the viral chromosomal DNA to replicate and would create two messenger RNAs. One would create an mRNA protein molecule that was representative of a particular gene—a gene He had selected from viewing the new DNA molecule. The other would produce the molecular form of the same gene. Knowing how to do this would help the cell workers to quickly match a gene in its protein form to its molecular form and speed up replacement of injured genes by using the molecular form.

To restart the process, He added RNA polymerase to the cell. This caused the promoter to take its place and instructed the polymerase to activate. Simultaneously the DNA strand started to unwind and elongate. The promoter is located about ten bases downstream of the initiation of the separation point. The mRNA strand is moving from five to three along the template strand of the DNA molecule. Only one strand will be used as a template and is called the template strand. The RNA polymerase moves along adding nucleotides to form the messenger RNA template which will be the section of the molecule that was requested. In molecular form it was two contiguous genes. This was done in the ribosome in Elohim's cell. If a ribosome was not present it would occur in the cytoplasm.

While this is happening, several polymerases were simultaneously translating the transcribed portions of the section of DNA that Elohim had selected. By adding amino acids chosen by using a reading frame that matched to specific codons, the template actively produces a ribbon that is a specific protein. The frame becomes a *table* that enables the matching of specific R groups on each amino acids to the correct nucleotide codon.

The reading frame permits sixty-four choices of matched codons and amino acids. Before the strand is released, the end of the strand sends a signal, if it is appropriate, to separate. This process can be RNA- or protein-related. It refers to two kinds of product. One product is molecular copy (digital copy) which can be produced in the ribosome and saved as a pre-mRNA template. The other is a strand of protein. If the second option is needed the process of loading the appropriate amino acids into the pre-RNA is followed.

Next, individual amino acids in the order they are needed are loaded onto transfer RNA (tRNA) units. They enter the ribosome and wait for the correct codon to arise. The synthesis of a protein cannot begin until it starts the codon process with methionine, which is found and linked to the first AUG codon in the messenger RNA template. This is a complex process, and it involves the use of several initiating factors. Since these more advanced features are universal for all eukaryotes they are often found in prokaryotes. It is important to be aware of them.

Once the AUG codon is found by initiation factors, the assembly of proteins using their messenger RNA template can start. Next, a methionine unit needs to be used to anchor the small ribosomal unit and locate the AUG codon. Finally, the large ribosomal unit is attached. Protein synthesis begins with the first amino acid which will be compatible with the AUG codon, followed by a string of amino acids that are complementary to the messenger RNA template. Transfer RNA will bring in the amino acids and drop them off on the messenger RNA template, then leave. The template is moving in a five-to-three direction. When the template is completed, it will be processed in the large ribosomal unit.

The large ribosomal unit has three sites. One site is an aminoacyl-binding site that binds to incoming charged aminoacyl-tRNAs. The second is a peptidyl P-site that binds charged transfer RNAs carrying amino acids that have formed peptide bonds with the growing polypeptide chain but have not yet dissociated from their transfer RNA. The third is an E-site that releases a dissociated transfer of RNAs that have released their amino acids and want to get recharged with waiting amino acids. Remember, all this information comes from research done by biologists. The naked eye cannot see the process that is occurring in the test tube.

This production of proteins by the messenger RNA method occurs during translation if a protein product has been requested. It involves the procedure described above. At the completion of the chemical process, a release signal after the messenger RNA was loaded was to request the protein form. The release is granted, and the protein is removed by a transfer protein and taken to the site in the cell that needs it. To get the molecular copy, a different signal is used and results in only the molecular copy being released. This is stored in the ribosome for use in forming messenger RNA templates in the future. During this intricate process the start point was bonded to formulated methionine producing tRNA. It enables the addition of formulated methionine to attach to the beginning of every polypeptide chain when it first appears.

The next step was to make sure that the template information was followed accurately. The RNA strand that is to be the protein product tracks over the messenger RNA template which has been bonded to the ribosome between the large and the small ribosomal units. The RNA strand starts to move over the messenger RNA step by step. Steps occur every time three codons begin to move in a three-to-five direction. As it did so, the first codon on the RNA strand that was tracking over it was registered as being okay. Simultaneously the strand moves in steps of three codons per step. As the RNA protein strand moves in a five-to-three direction, the messenger RNA moves in a three-to-five prime becoming a completed protein product of the gene which is delivered to wherever it is needed. If the direction is slow and the protein is not used quickly, it will deteriorate. It should be noted that until the amino acids are added the molecule is simply a pre-mRNA and is not a protein, so it is less susceptible to weakening.

These steps have been confirmed by scientists using a variety of techniques that detect the intermolecular reactions of nucleic acids. The various enzymes and cellular proteins act on their own; it is astonishing that, despite the complexity of this process, errors are rare.

The extra codon may cause an extra molecule but does not affect anything else. And it must work because since the beginning of life all living things have used this code without consulting any human. And they have flourished.

The enzymes involved in these processes are amazing. They collaborate sequentially with each step and with very few errors. They do not wait to receive messages from Elohim and are self-sufficient. A brief description of some of the players, which are all proteins, is below.

The most well-known protein is RNA polymerase I. It is a nucleolar enzyme and is responsible for the creation of some structural products that are processed into small rRNA molecules which help with building ribosomes in the cell. They are not involved in protein manufacturing but are considered important in building structural features in the form of rRNAs that are essential for translation in the ribosomes. It is the only site where this can be done in prokaryotes. These RNA polymerases have several responsibilities in the functions of the ribosomal RNAs. Enzyme I's responsibility is to make sure they are not having any problems.

Ribosomal polymerase II, the next enzyme, resides in the nucleus. It produces protein-coding nuclear pre-RNA in eukaryotes. Most messenger RNAs that the ribosomes receive are not true proteins. They are pre-RNA molecules that are always subjected to intense processing between transcription and translation. Polymerase II is responsible for transcription of most of these.

Polymerase III is responsible for most translations. Located in the nucleus it is more involved in working with small structural molecules like pre-transfer RNAs, or pre-RNAs. It specializes in working with translation of the messenger RNAs of large molecules. They work as adapter molecules that assist with loading or linking small protein fragments with the main molecule when a molecular copy is being produced. All this happens and has been happening automatically for billions of years without any fallout that we know of. If that is not a miracle, I do not know what is.

Elohim remembered the need for available energy sources once the cell became active. He added a simple metabolic cycle called the citric acid cycle into the cytosol of the cell. He also added some ATP, and GTP, another phosphorylation-related energy molecule, enclosed in a vesicle. His next action was to transfer the prototype prokaryote cell to all landmasses and the oceans. He made sure they lay in environments suitable for life. It would not be long before He would need the cells. All creatures alive through the ages are supposed to have been derived from this last universal cell ancestor. They all carry elaborate defensive and quality control mechanisms within their genomes. This extremely detailed and logical process is only one of many complex processes that govern the behavior of matter. It cannot be attributed to nature.

Chapter Six

Meiosis and Sexual Reproduction; Body Plans;

First Eukaryotic Cell Ancestor (FECA)

Around two billion years ago, Elohim activated the cells He had in place. Most of them successfully made the transition to active living cells. Elohim added more sophisticated energy cycles to make possible the active cell division that occurred. By two and half billion years, the prokaryotic cell of bacteria had increased in complexity. During the process of cellular merging, some cells had acquired a double-stranded DNA chromosome enclosed within a nuclear membrane. The nucleus was surrounded by cytoplasm in which floated several ribosomes containing RNA. They are *centers* performing protein replication. For protection, the cytoplasm is enclosed in three layers, a cell membrane made of phospholipid, which is surrounded by a capsule made of peptidoglycan, and is enclosed in a capsule made of lipopolysaccharide. Many cells contain small strands of DNA, called phage particles, in the cytoplasm. Elohim decided to use one of these cells to set in motion the formation of the first eukaryotic cell.

Elohim saw that left on its own, the development of the first eukaryotic cell would probably never occur or take a very long time. It would involve many changes to become a simple prokaryotic cell. Meiosis is the process for mixing chromosomes in male and female gametes. It is a major cause of variance since it causes changes in chromosomes every time an embryo is created using the process of meiosis. New members added to a clan also introduce more variances in the clan. Sexual reproduction occurs when a species evolves the anatomical structures and the physiological behavior that typify the mating of male and female gametes. The mechanism of accomplishing this event varies from species to species. Once an egg is penetrated by sperm it forms an embryo which grows into a new member of the species. That is simple enough. What is complicated is the process by which the sperm (male gamete) and the oocyte (female gamete) are formed. When the time is right the female will create oocytes which will remain for two or three days in the stroma of the ovary. During this time a complex process begins to reduce the total chromosome count in an egg from twenty-three paired chromosomes to twenty-three single chromatids. This is called meiosis. In the resting maternal cell, called an oocyte, chromosomes are present as single chromatids enclosed in a nuclear membrane.

The first thing to happen is that the chromatids in a maternal active egg cell replicate causing their number to double. Next, homologous (similar) chromosomes join into twenty-three pairs of chromatid strands. These are two identical strands that form a chromosome. During cell division the chromosome divides into two chromatids. In the next step, called prophase, homologous chromosomes connect with each other along their length and, in the process, small segments of the chromosomes are exchanged between them. That is the simple part of sexual reproduction and the most important. The problem is that the egg or the sperm at this point each have twenty-three new paired chromosomes. When they form an embryo, it will have forty-six pairs of chromosomes, which is twice the number needed. Elohim's solution for this was the card trick of meiosis. This process of merging the chromosomes of two individuals into one set is complex. It is different from mitosis, which is simple cell division with both the mother cell and the daughter cells having the same number of chromosomes at the end. Mitosis is used only for making new cells from old ones in the body. Elohim had introduced this as soon as enough cell lines had formed from LUCA (last universal common ancestor).

Meiosis has two purposes. First, it provides a mechanism for genes to mix which produces an entirely different individual compared to the parents. This greatly increases the chance of new traits. The second purpose is to produce eggs (or sperm) that contain a single set of twenty-three chromatids instead of twenty-three paired chromosomes (each with twenty-three chromatids). This is accomplished by a special type of cell division. First the replication of chromatids occurs in the mother cell. This doubles the number of chromatids to forty-six and the number of chromatids when these same cells split in two will be twenty-three.

The same cells when chromosomes split in two will be forty-six chromatids. One to two and two to four chromatids then form units of four chromatids linked together by centromeres called tetrads. They connect with each other longitudinally through a process called synapsis. Next, they engage in a new step during which the homologous (similar) chromosomes exchange segments. They exchange small segments of themselves by connecting at the ends and breaking off. The segments exchanged are random. This produces two completely different sets of chromosomes in each cell compared to the initial maternal cell. Next, the two sets of chromosomes line up at the equator of the nucleus with matching homologous chromosomes together. Once this is completed, meiotic division occurs. The cell divides into two and each cell receives one of the two homologous chromosomes. Each daughter cell in meiosis contains twenty-three pairs of chromosomes. The new chromosomes are different from the mother's chromosomes because of synapsis and the crossover of the chromatids. The next meiosis step is like mitosis. The twenty-three chromosomes in each of the two cells separate into forty-six chromatids each. Each cell then divides into two cells producing four new daughter cells each with twenty-three chromatids. These four cells are the ovarian eggs. A normal cell would have two chromatids per chromosome which is forty-six chromatids. The four ova that were produced have half the number of chromatids, with twenty-three chromatids in each of the four ova (eggs). A similar process will occur in the male sperm which will have twenty-three chromatids. The embryo will receive half its genes from the mother and half from the father and will therefore revert to two sets of twenty-three chromatids, which is twenty-three chromosomes (a normal twenty-three chromosomes with forty-six chromatids).

A similar process occurs in the testicles of the male, rendering haploid gametes with twenty-three chromatids each. When fertilization occurs, the new embryo will have a complete set of diploid chromosomes. Because of the exchange of DNA segments, they will be different from either parent. By this amazing process Elohim instantly introduced meiosis and sexual reproduction into living things. The initial mechanisms to integrate the process were primitive in eukaryotes in that they did not yet have bodies complex enough to have organs of sexual reproduction in place.

As the complexity of organisms increased, Elohim produced creatures such as earthworms that have both male and female structures in one body and both produce eggs and sperm. By attaching to each region of its clitellum they exchange sperm and save them to inoculate their eggs. The pull-apart and the process of egg production begins and is followed by fertilization from the other worm's sperm. There, cross-fertilization occurs. The clitellum will thicken and become a sac which eventually ruptures to release the thin white worms that will grow into adult earthworms. Each worm could, if it wanted, clone itself by using its own sperm. In general, they do this because they are aware that it results in unhealthy progeny (they are not the healthiest progeny products). Both males and females have a saddle-like structure called the clitellum from where they can both exchange sperm. Each earthworm can produce both eggs and sperm. When this happens, it is called hermaphroditism. Many fishes also had this organization. Over time, creatures evolved with separate genders and increased in overall complexity. Rapid improvement in morphologic features occurred. From the time of sexual reproduction in its simple form, complex morphological improvement and complex biochemistry in the form of hormones has totally changed the efficiency and productivity of sexual reproduction. This was much more efficient and effective than binary fission in producing new prototypes. The mechanism of how this amazing transformation occurred is not clear to science. Another unexplained phenomenon is the rate at which speciation and new body plans occurred. In the space of twenty million years in the Cambrian Era, twenty-four of the body plans that now exist evolved. The remaining twelve were introduced slowly in the late Cenozoic and, as of now in the twenty-first century, we have a total of thirty-six body plans.

Elohim, for His part, was pleased with the result of His work. He knew that it was not perfect, and that occasionally errors in genetic sequences would occur. Also, in the right setting, with adequate nutrition and normal survival, it would be easy for Him to manipulate the genetic sequences to produce specific structural or metabolic improvements. This option would facilitate the process of the creation of Homo sapiens when the time came. Having introduced mitosis, meiosis, and simple sexual reproduction, Elohim felt ready to craft the final eukaryotic cell.

Elohim did not need to wait for stepwise evolution to produce a functioning eukaryotic cell. He gathered all the important structures and put them in place in a cell with all the biochemical materials like enzymes, amino acids, and proteins. He added all the essential structures that would make a functional eukaryote. It contained many new structures. The first big difference was that DNA chromosomes were contained in a nuclear envelope instead of lying free in the center of the cell. There were also more chromosomes. He had programed them with His own expertise to have all the information that had been in LUCA. This was by design. In case the LUCA cells became extinct, their historical DNA would be present in the new eukaryotes. The nucleus had a nucleolus which would orchestrate the process of meiosis. The double-stranded DNA came as three chromosomes, which were programmed by Elohim to form this new cell which would be called FECA (first eukaryotic cell ancestor). DNA was a symbiosis of bacterial, archaeal, and viral genomes. He used DNA strands and reorganized them to fit the needs of a basic eukaryotic cell. He would program them for some of the many new processes that the cell would be capable of doing. As it evolved, the cell would learn how to change genes by adding appropriate DNA strands to produce cellular effects.

He especially emphasized the creation of His control genes that could manufacture new genes for each cellular activity. He selected genes which had these talents and coached them in a process which involved using existing genes or creating new ones by manipulating them so they could turn them into control genes. They used eukaryotic, bacterial, and viral genes to form a symbiotic mixture that could be used by cells to create new chromosomes. Next, He added two new structures in the cell. One was the endoplasmic reticulum (ER) and the other the mitochondrion. The former is a membrane-bound space which is folded into compartments and is contained in the cytoplasm. There is a smooth endoplasmic reticulum (SER) and a rough endoplasmic reticulum (RER). The RER is studded with ribosomes which is involved in a stage of protein synthesis, specifically the folding and modification of protein for transport. The SER is involved with lipid and steroid synthesis and carbohydrate metabolism and detoxification. In the early cell, the main function of the standalone ribosome was to accomplish the translation of messenger RNA and to generate phospholipid for cell membrane replacement. So, the mitochondrial ribosome is different.

The mitochondrion houses an enzyme-dependent energy-production factory. Ribosomes are present and contain ribosomal RNA which in eukaryotes is involved in protein synthesis within the mitochondrion as well. The new eukaryotic cell was capable of both symbiosis and phagocytosis. Absorbed material like bacteria, viruses, or cell debris are engulfed by lysosomes which then digest them. The unused byproducts are ejected by the cell. The proteins are retained by the cell and used for whatever they are needed. In addition, He also created extra genes called homeobox (Hox) genes. These genes carried genetic information for a body plan that could be passed on to the cell's progeny and continued through generations. It details and stores the embryonic timeline for each species for which it was created. Elohim decided that He would have about thirty body plans and would use some plans for more than one species. His main goal was to quickly and safely produce an imprint that would start the evolution of eukaryotic cells and produce many creatures with different body plans. When the eukaryote prototype was created to His satisfaction, Elohim released it into a location where it would proliferate. This was the LECA (last eukaryotic cell ancestor), the ancestor of all animals. All animals from then on had the features seen in the LECA and Elohim added improvements as time went on. Elohim was confident in His creation of the first eukaryotic cell and His body plans. The many biochemical and anatomic changes He introduced caused a population explosion of eukaryotes. Initially it was in the form of binary fission because, although now programmed for mitosis, the memory of binary fission remained. Elohim did not mind this as it would be good to increase the population by a short run of binary fission. Every cell would produce two cells, and two cells would produce four cells. It would make sure that the new population would not fade away and become extinct because they had not been tested for the new features He introduced. However, there would be no variation and, in the end, speciation would be limited.

Once meiosis was established, the organisms were able to rapidly increase in complexity and variety. Viral and bacterial/archaeal DNA continued to blend the same way they did in prokaryotes and new structures were formed, such as flagella and pili. Some of these structures were part of the body plan mentioned earlier. All animals that were produced after this time were genetically connected to this ancestor. Over the next twenty million years, many new species formed and rapidly multiplied. While monitoring the results of the introduction of the LECA, Elohim made a discovery. In an area we define geologically as the Ediacaran Ocean, He discovered an unusual activity. The older biotas in this area had died and were replaced by an entirely new population of creatures. They were significantly larger than the ones that became extinct. Soon they organized into three groups.

The first, still unexplained, was the Ediacaran biota found in certain types of environments that occupied the late Proterozoic. Fossils of Ediacaran flora were found as imprints in rock layers from the Precambrian in Australia, Canada, and South China. They appear to have evolved possibly forty million years before the Cambrian explosion occurred and were far more complex than the eukaryotes that developed in the Cambrian. Imaginary three-dimensional (3D) reproductions of them have been published, showing complex creatures with advanced morphologies which are far more advanced than any eukaryotes in the Cambrian seas. Since they have been found as imprints, it takes some creativity by artists and paleontologists to produce the three-dimensional representations. Some of them are shown in detail and in a 3D perspective, so considerable artistic privilege is exercised. We still do not know much about them, not even identifying them in any kind of classification of life forms that we have today. We do know their fossils have been found at several sites which have similar geologic characteristics. These are present as imprints in sandstone rocks and happened when they died, because they were in shallow waters with sandy bottoms. When their bodies dropped to the ocean floor and became embedded in the sand, thus becoming sandstone rocks.

Next, an unusual and sudden evolutionary activity happened. A vast number of new organisms or species evolved within five to twenty years in the Cambrian ocean. The numbers represent the possible years it took to produce the exceptionally large number of variations with mostly new species. Both are possible time intervals for the event and both are hard to explain. There was no extinction event to create niches with scanty resources to create the scenario of a post-extinction burst of new species. This is the setting when Darwinian-type speciation is most likely. This event occurred around 580 million years ago. We call this phenomenon the Cambrian explosion, and it is just that—the sudden evolution of thousands of species. All of these were eukaryotes, descendants of the LECA. They did not show the extreme variation seen in the Ediacaran but this degree of evolutionary advancement with numerous new species and advancement in complexities of eukaryotes was unprecedented. Elohim's LECA had done its work.

Elohim decided to be proactive in introducing some controls into the process of evolution by natural selection. He would introduce a control system in the form of regulatory genes that would automatically monitor each new genetic feature that was presented to the cell, as an addition to it. He had parameters for assessing how much burden the trait had for the species. If the burden or risk was high, the regulator genes slowed down the process of accepting the new feature by establishing a constraint on the feature. These regulatory genes organizing the constraints worked to move things toward rapid acceptance or slow progression depending on what directives Elohim prescribed for them. He was explicit enough in His regulations that the regulatory genes rarely needed to refer a situation to Him. Somehow, they were able to identify if the effects of the new genetic trait were harmful and prevent further progression of detrimental results. On the other hand, if it was deemed beneficial the acceptance of the trait was facilitated.

The constraints Elohim created for the thirty-six body plans had specific commands. Processes that were essential for survival would be severely constrained and it would be unlikely that any new traits would be allowed in that setting. Elohim made sure that He was made aware of any unusual happenings. Appropriate messaging systems were designed by Him for each body area using a baseline criterion of risk or burden for that area so that there was a balance between development and constraint and a constant awareness of the burden displayed by the trait. If there were malfunctions of His coding, Elohim arranged for a means for the cell to reach Him via dark energy. Body systems that were exempt from being considered as candidates were defined by Him. These would be the DNA systems governing replication of important proteins, or systems that were part of the regulatory genes systems that was the Hox-TALE combo. They would not interfere without His involvement. In these systems, unfamiliar traits were rapidly extruded. The measures Elohim put in place were mainly for prevention of negative results from imported trait adoption. Having evolved a limited number of body plans with a finely tuned set of regulatory genes specific for each of them, Elohim felt the evolutionary process was in safe hands. He did not intend to produce any more body plans. Evolution in existing species would be restricted to the phylum level and below. Speciation would continue and produce variations so that His creations would not get monotonous.

Research in this area may give us some better understanding of body plans and first began in the 1970s. Rupert Riedl, born in Vienna, Austria, was a marine biologist who became interested in evolutionary biology. In 1968, he became a professor at the University of North Carolina. At that time most of the evolution research being conducted was related to population genetics with an emphasis on producing accurate taxonomy and phylogenies for following and possibly speeding up the evolutionary process. He felt a broader approach should be researched along with other aspects of biological investigation to determine how evolution sped up. The main example of this was the Cambrian explosion which led to the appearance of all body plans existing today within a period of five to twenty years. With its slow one-gene-at-a-time process, Darwinian evolution by adaptive advantage did not seem to be the process at work here because it was too slow. Also, random genetic mutations were much more likely to be deleterious and cause the destruction of an animal. And the new families of animals in the Cambrian explosion did not display a marked increase in mutations.

Scientists today are still unable to explain the existence and formation of body plans. They have conducted a variety of biological experiments to confirm their existence and have an elaborate explanation of how the messages are sent to genetic regulatory networks (GRNs) that they suspect are present. Researchers have looked for and found theoretical communication techniques named by electrical connector terms. These include kernels, plug-ins, input/output switches, and gene batteries. These connectors work around the GRNs. They hope that their theoretical roadmap will evolve further, and that molecular, morphological, and computer-mapping analyses will give them answers to the two questions that have not been answered. The first: "How did animal body plans first arise?" The second: "How have animal body plans remained essentially unchanged since their Precambrian beginnings six hundred million years ago?"

Elohim's next move was to increase the genetic mutation rate slightly and see what happened. Until the earth was populated with most of the phyla of living creatures that He planned, evolution would have to be accelerated. If all went well and He achieved a faster turnaround on new species or individual traits, He would continue the mutation rate until the body plans number of thirty-six was fulfilled. He would leave further changes to Darwinian evolution. The slow progressive change would continue and produce perhaps unusual advancements in evolution which would gradually freshen up the population as the earth aged.

Chapter Seven

The Cambrian Explosion of Life

600 mya–543 mya

Six hundred thousand million years ago, Elohim surveyed the results of His efforts to organize early life. He had let nature take its course after introducing several new structures and processes into a bacterial cell. The process of evolution that He had set in place was improving the structure and survivability of all families of creatures, resulting in new and interesting species. Life so far existed only in the oceans. The earth was thawing from a snowball earth, the result of an ice age caused by the increase in organisms that survived by getting energy and nutrition from the chemical process of photosynthesis. The release of large amounts of oxygen as a byproduct of photosynthesis caused the oxidation of the greenhouse gas methane and as a result temperatures dropped. Methane was the main greenhouse gas at that time, which meant it was the main ingredient that acted as a blanket for the earth. As the earth gradually lost heat from its surface, ice formed and the earth was covered in ice both on land and in the oceans. Life remained in the depths of the oceans where the methanogens lived. They continued to make methane.

Another factor that reversed the process was the release of carbon dioxide from dolomite rocks that were present in many areas. Dolomite is a type of calcium carbonate. When exposed to acidic substances, it breaks down and produces carbon dioxide (CO2).

Another path to increasing atmospheric CO2 was the natural result of the melting of the glaciers in the middle range of altitudes leaving sediment that was rich in nutrients. Phosphorus was one such source. Combining with other chemicals in the sludge it could produce nutrients that methanogens need. A surge in population of methanogens occurred, causing high levels of methane rapidly filling the atmosphere. Added to the surge of carbon dioxide from reactions with carbonates, the greenhouse gases rapidly increased. It has been determined by geological techniques that the carbon dioxide level at that time in the earth's existence were three hundred times those present today.

Both gases rapidly released into the atmosphere as the crust was exposed to the air in midlatitudes. Wind blew dust from the glacier ruts on the iced areas and they dissolved faster. We cannot really predict the chances that a snowball earth could continue to chill and become progressively frozen to an inactive entity. Scientists say that it is a possibility. It did not happen to our earth, and we are grateful for that. Elohim, of course, witnessed the whole event and let things go their way. It was a thrilling experience for Him. This was a good time to experience it when there were only early single-celled organisms around. Most of them sank into warmer depths in the ocean and when temperatures warmed up, they profited from the large amount of glacier sediment packed with nutrients that washed into their refuges. If things looked bad enough to threaten the formation of living things, He would immediately have intervened. By the time the Cambrian started, the earth had recovered from its snowball stage. The climate was balmy and no threatening extinction events were imminent.

Volcanic events introduced more methane into the atmosphere and a warming trend began. The balmy conditions seemed an ideal time to increase the population of the oceans. He encouraged the growth of large numbers of tiny creatures that multiplied in sheets that we call plankton. They would provide food for other creatures. They drifted in ocean currents, enabling the development of very active ecosystems across most of the warmer areas of the planet. He also nudged minor changes in some of these systems and watched, fascinated as diverse life forms came to be. This occurred in a very short interval compared to the lifetime of our earth.

According to the geologic time scale (GTS), the Cambrian Period extends from 543 mya to 248 mya. By six hundred million years ago, four billion years after the earth came into being, single-celled organisms had evolved into multi-cellular organisms with internal organs used for food processing, as well as simple types with eyes and protective hard skeletons. Numerous metazoan creatures emerged. The first sea snails appeared and there were early versions of crustaceans and arthropods or insect-like creatures.

Elohim observed the proliferation of life forms in the warm Cambrian seas. His attention was drawn back to the extraordinary collection of sea life that was occurring in certain areas of the oceans. In one area in an ocean, called the Ediacaran, He found an amazing collection of sea creatures. Their morphology was complex and bizarre by existing expectations for that time.

They ranged in size from one eighth of an inch to several inches and a few of them resembled many of the families of plants and animals alive today. Elohim studied them with fascination, impressed at what His DNA molecule had achieved in such a short time on the geologic scale. He knew that most families living in the Cambrian would become extinct as a result of devastating destruction by ecological disasters of one kind or another. Before this came to pass, He needed to nudge into existence creatures that would depend on their survival by consuming other species, known as animals.

The energy derived from photosynthesis by algae and later plants would not support the high levels of activity of more complex organisms. While some of their nutritional needs could be met by consuming plant products, complex locomotion and sexual reproduction, which results in the bearing and raising of offspring, would require a more complex diet, one with complex animal proteins and fats. He hoped the advanced organisms He had in the Ediacaran beds would produce some predatory species.

Elohim also saw that at the rate life was advancing there would soon be a need for many new biochemical cycles that would control the production and multiplication of the complex cells of muscle, bone, and the various other organs that would evolve. He proceeded to introduce biological metabolic cycles that integrated the ATP/NADH energy-producing cycles that are found in the Calvin cycle in prokaryotes. As creatures became more complex, He was ready with several more complex metabolic cycles, some specific for different organs. He thought about the types of the possibilities, and they were endless. He had already experienced many families of animals in previous universes. He had plenty of time to think about it. He would watch and see what chance may bring Him.

As organisms increased in complexity, they would need a homeostatic linking of all chemical processes, and He looked forward to establishing this system with considerable excitement. Creation with every detail addressed was one of His favorite activities. This new world looked like it would keep Him busy.

His plan was to proceed with the production of numerous species built around a limited number of body plans. He did not want to wait for the random occurrence of genetic mutations spaced thousands of years apart to produce successive evolutionary changes. He would go directly to the existing cellular mechanisms and change the instructions within the growing embryo so that any new features He wanted the animal to have could be achieved by simply changing the body plan and adding DNA manipulation to create any new features. He liked that idea. It would give Him a chance to introduce a variety of unique features.

Elohim could recall any detail about any creature that He had already seen or create new ones from His own imagination. He was more worried about whether He could imagine new species. His concern was whether the thirty-six body plans that He had already started would be enough to fit in all species. Each species needed their own body plan. But in the interest of keeping the number of body plans at a reasonable number He had settled on thirty-six. To fit His creation in the major species groups, He calculated that He needed thirty-six embryonic body plans.

Embryological body plans map the different stages that a particular animal's embryo goes through while in the womb. The changes occur as the embryo grows to maturation. Different parts of the body evolve at different times. Elohim was able, at crucial times, to affect embryonic changes as a body part formed. By doing so, He could change what the animal looked like or add a new feature. It takes skill and imagination to do this effectively, as it resembles artwork. Elohim had been designing different mammals and was enjoying it. Creativity was a much more effective way of increasing variation. He knew the genetic workings of cell differentiation and could encode multiple variations that are mainly aesthetic and not likely to affect vital systems. His ability to control a system enabled Him to orchestrate the appearance of new ones with ease. In fact, as the embryo developed, He was able to introduce special features by adding information at the right time.

His final decision had been to try to use one body plan for two embryos. By the introduction of specific genes into each embryo He could produce interesting but not essential features into the prototype of the creature He was producing. For instance, tails that were needed for two entirely different species within the same body plan appeared to be identical in the early stages but ended up being the usual shape when they matured. This enabled Him to program the development of two species in one body plan. Two creatures who have similar-looking early embryonal appearances would develop simultaneously and emerge representative of whatever species they were designed to be.

He started with thirty-six plans and soon realized there would be more than one species with the same body plan. On the other hand, with species extinctions in large numbers during disasters, He would regain body plans regularly. He calculated that the number of plans would work as long as He could have two or three creatures in the same body plan. If needed, He could produce new body plans.

Eventually there would be embryological roadmaps for about thirty-six body plans. Elohim planned to allow the body plans to evolve close to the time when He would begin working on man. Once He started forming man, He would stop creating new species and let the slow process of natural selection determine what evolved.

Elohim continued His creation of new mammals. He put in place biochemical systems both in the intestinal tract and in various organs to establish the bioavailability of proteins. This is a simplified version of this process and gives you an idea of how detailed Elohim was in creating man. He introduced each process and then made sure it would be monitored and maintained. Looking in detail at the ingenuity and resilience of each species of current biologic creatures is amazing.

To jumpstart the evolution of new life forms, Elohim decided to take advantage of the pleasant climate during an interglacial period to increase the variety of life forms. He slightly increased the basal mutation rate of living cells, which He had preset at a low steady rate. The result was an explosion of life with all kinds of phylogenies. Within a short five to twenty years, a blink in the five hundred million years to follow, two-thirds of the body plans for living things that are precursors of ones that exist today evolved. He knew that many species would become extinct in major extinction events. Their body plan would be removed when this happened, making room for a new one.

By rapidly introducing new phylogenies He had reduced the time needed for the specialization of more complex forms and a great variety of intermediate creatures evolved. These included corals, brachiopods, and trilobites, many of which later became extinct. The most important of these from an earthly standpoint were the varieties of fish. Other than plants and bacteria and archaea, there were no creatures on land. Most life was still in the oceans.

The biggest collection of fossils of the late Proterozoic Era comes from a geological find in the Burgess Shale of Canada, the Ediacaran Hills in Australia, and in the South China sea. These areas were under water years before the late Proterozoic. The Ediacaran flora and fauna are found in many shapes, some of them complex, and range in size from one-eighth of an inch to two inches. They lived on the sea floor in shallow waters and left imprints on the soft seabed when they died. These were preserved and have given modern-era scientists a good idea about what they looked like. Artists have created three-dimensional models based on the imprints but have no idea what internal structures they possessed. Researchers do not know whether they should be classified as plants or animals. Also, they do not know what type of metabolic processes they possessed. Unfortunately, the number of fossils found overall in the late Cambrian Era was low and because of this, scientists do not have much information of what they evolved into. Although most of them became extinct, they bequeathed segments of body plans to future genera and species that were incorporated one feature at a time.

Elohim, looking at the life forms, saw the living precursors of our current-day arrow worms represented by a fossil we call Protohertzina, and the snail-like creature Aldanella representing future mollusks. There were enigmatic creatures like Opabinia that looked like a small turtle with teeth. There were many strange-looking creatures resembling scorpions. Trilobites and Marella splendens, an exotic spider-like creature representing the future insect world, were examples of these. The majority of the creatures were less than two inches long. There were a few that were larger, one of them being the Anomalocaris, a three-foot-long segmented predator with protuberant eyes and sharp mouthparts. It was probably one of the top predators of the day. Most of these creatures were invertebrates probably belonging to the phylum Arthropoda. Today, arthropods include crustaceans, arachnids, a great variety of the class Insecta, and many types of worms. Elohim was amused at the attempts of these tiny creatures to intimidate other species, as the transition from plants to animals that fed on other creatures proceeded. He realized that some of the new creatures looked familiar, and He remembered the very advanced creatures He had seen in the late Proterozoic, as He gave them a helping hand by adding some more sophisticated advanced features. He realized that it was time to regulate the process of evolution to fit within certain guidelines rather than let things come to be just because they were interesting. He knew that there needed to be a fair balance between predators and prey and that having creatures that only consumed would not work. The earth would soon exhaust its raw materials to make living tissue.

So, He established a hierarchy of plants and animals that would use the environment in which they lived to find food and shelter. He gave those that were weak protective features like shells. But He also introduced simple predators that would feed on other animals. Life would be forever marked by a struggle for survival. Weaker creatures would become food for stronger animals, thus establishing a cycle of reusing nutrients. He also foresaw that size alone would not be the deciding factor in survival. In fact, arthropods, or insects, because of their resilience in a variety of environments, are the largest phylum of the animal kingdom and 84 percent of animals belong to this phylum.

The sudden massive explosion of speciation within twenty million years in the beginning of the Cambrian is not a mystery. It was Elohim at work evolving life-forms at a reasonable pace and controlling the forms that survived so that there would be the right mix of creatures for the resources available. Variety would be provided by new species evolving, instead of new genera.

This picture does not fit in with a divergence by natural selection and adaptation. A short span of time, like twenty million years (five million years according to some), is not enough to produce the extent of variation that was seen. Natural selection with increased speciation is fastest in situations where a great loss of species, or even families, has occurred. A few survive and will be resistant to the offensive event. They will produce a new clade with variation and new species forming and benefit the next time they encounter the destructive agent. They will be "fitter" because of it but with a large loss of family members. There was no evidence of such a scenario in the Cambrian explosion.

Most natural mutations are destructive rather than constructive. The cumulative process of forming a species with a natural process should produce little divergence in a short time. This process of immense speciation in twenty million years does not coincide with the normal natural process. It was Elohim initiating new life by directly making various changes by manipulating genes. It was controlled creation at work. It fits better with the theory of punctuated equilibrium put forward by Stephen J. Gould and Niles Eldredge, who suggested that evolution occurred in surges against a background of slow evolutionary change. The Cambrian was a time when climate was ideal, and no extinction-type events occurred. There was no isolation of small groups of a species. It was not an environment that would expect rapid speciation. Also, for rapid speciation to occur within five million or even twenty million years was very unusual. We do not have a scientific explanation to explain it.

Elohim had given much thought to the Ediacaran fauna. He was not certain how creatures with such out-of-phase complexity would fit in with the simple, mostly marine creatures that were beginning to evolve. If they continued to speciate randomly in such dramatic ways, the disordered speciation could lead to conflict between the two levels of new life. He decided that He had produced them prematurely and would have to remove some of them. He selected multiple species with very complicated body plans and moved them from the current habitats they inhabited so that they would not be able to transmit their genes to other species. He isolated them in shallow ocean in coastal areas. He produced deep ponds surrounded by rocks in each of the areas where He had initiated the Ediacaran flora. Then, He made sure that the ocean bed was sandy so that when they died, they would drift downward to the ocean floor and leave imprints that would document their existence. Elohim also gave them a shorter lifespan and hoped they would reduce in numbers and disappear. He would need to keep checking on them, as the fauna were accustomed to the balmy Cambrian climate and probably would not survive a cold spell. Until then, they would enjoy their lives.

Chapter Eight

Ordovician Mass Extinction:

Satan's Attempt to Destroy the Earth

490 mya–443 mya

The Ordovician Period followed the Cambrian Period and accounts for the next forty million years of the earth's history. Because of ongoing tectonic activity and the resultant outgassing in the ocean beds, greenhouse conditions prevailed with high atmospheric content of carbon dioxide, methane, and water vapor. The oceans were warm, and this led to continued diversity in the species of animals, especially arthropods. Arthropods at this time began to migrate to the existing landmasses, the largest of which was the continent of Gondwana. Polar glaciers were absent, leading to high ocean levels. In this warm climate, a milieu of coral reefs grew, providing the potential for new ecosystems and increasing the proportion of landmass compared to ocean.

During this period many species of fish populated the oceans. The early fish were flat, jawless, translucent creatures without fins, an intestinal tract, or bony skeletons. Elohim took the opportunity to start a new line of fish that would have notochords, bony skeletons, and hinged jaws. Their creation would be an important milestone for the evolutionary process. They were the starting point for the ancestors of vertebrate animals from which man would eventually evolve.

Besides being essential to the development of higher vertebrates, these features also gave them new and improved survival advantages in the environment of the early Ordovician. The jaw, in addition to allowing the biting of targeted prey, also improved respiration. The gut provided the processing of ingested food to enable a higher form of metabolism, and the notochord allowed the development of an early nervous system. They were called chordates and were the precursors of the vertebrates that would evolve later. With these advantages, the new breeds of fish dominated the Ordovician environment resulting in the extinction of jawless fishes.

Haikouichthys is the oldest specimen of a chordate fish which has been found fossilized in marine sediments in Chengxiang, China. Fortunately, Elohim knew the future and made sure that He monitored its life and other similar species, so that they never came close to extinction. If its line had ended with extinction, human beings would not exist. The species was at the roots of our evolutionary tree.

Elohim continued to encourage the survival of the chordate fish which over time evolved into coelacanths. These fish were called lobe-finned fish because their fins were soft and rounded. They would eventually evolve into the feet of tetrapods. Many species of these had inflatable air bladders that obtained air by gulping movements in their pharynx. This gave them buoyancy when they needed it. Much later they would develop into lungs in the new tetrapods and amphibians.

Elohim pictured these early body structures growing in sophistication and size through many cycles of evolution and extinction. He felt a sense of loss when He thought of so many species becoming extinct but knew that they would be reintegrated into His matrix and appear again as other species. Once all major groups of animals evolved, He intended to gradually increase brain size, working toward a large brain and sophisticated spinal cord of His eventual sentient being, man. While He waited for that event, He took pleasure in observing and editing the morphologies of the numerous exotic life forms that filled the oceans of the Cambrian. During the Cambrian Era, all eukaryotic life existed only in the oceans.

The explosion of life continued into the Ordovician. The already existing families of organisms in the warm oceans increased in numbers. Large coral reefs developed. Satan watched this proliferation of life with great displeasure. He knew that it would be a while before Elohim would start working on the creation of man. However, he was envious of the enjoyment that Elohim was receiving from the multitude of creatures around Him. He watched Elohim look tenderly on the weird creatures of the early Cambrian and was envious of Him. Besides loving the new creations, Elohim took great pleasure in bringing into being one new creature after another. Secretly, Satan wished he had this ability, and he decided that he would play spoilsport and create a setback to Elohim's creations. He was determined to destroy the prolific life he observed in the Cambrian oceans.

Satan discussed his feeling with Beelzebub and asked his advice. Beelzebub was thrilled to help him out. Locked into Elohim's matrix, it was difficult for him to effect changes against living creatures. He did, however, know a lot about the science of how planets and solar systems behaved. He noticed that the earth was approaching the beginning of two periodic events that are now known as the Milankovitch cycles, named after the man who first described them.

The shape of the earth's orbit around the sun varied from round to an eccentric ellipsoid over the course of about a hundred years. The tilt of the earth's axis also varies from 21.5 to 24.5 degrees every forty thousand years. Currently, it is in the middle of the range at 23.5 degrees. A third variation in the movement of the earth is called precession and describes the wobbling of the earth's axis as it orbits the sun. The physics and the frequency of these movements would at times coincide with other events like volcanic eruptions. Two effects could then combine to cause serious changes in the earth's climate. If the atmosphere filled with volcanic dust and the event coincided with much of the earth being tilted away from the sun, cooling would occur and glaciers would form, resulting in long winters. Sea levels would be low due to water being trapped in ice sheets and many ocean creatures living in shallow waters would die.

The opposite happens if half the earth is tilted toward the sun for longer periods. Eruption of gases like methane and carbon dioxide produce a greenhouse effect, causing warming of the earth and an arid climate. Complicating things is the occurrence of periodic ice ages where longstanding cooling would cause glaciers not only at the poles but in other land areas. These are usually followed by an interglacial period when the climate is mild, and ice sheets melt. This complicated process, not experienced in kind by our short-lived human race, is still not well understood. Theoretically, there would clearly be devastating effects if the cycles peaked at the same time as other events like a volcanic eruption.

We are currently in an ice age which began only two million years ago. Icing in an ice age is not continuous. There are periods of thawing called interglacials, followed by a period of re-icing called a glacial. Currently, the earth is in an interglacial. Unlike past ice ages, which are often the results of natural events, our interglacial is unfortunately affected by manmade factors that are causing the warming of the atmosphere without the presence of a Milankovitch event. High levels of greenhouse gases from fossil fuels are warming the atmosphere and, unless this trend changes, we may not have the advantage of a glacial event that can rebuild the glaciers. This means that rising sea levels will stay with us and cause serious problems for cities and towns along coastal areas. Beelzebub explained these effects to Satan and suggested ways that could be used to induce catastrophic effects on the earth.

Elohim was aware of the effects of the Milankovitch cycles and was not concerned. He would work around these natural events to make sure that His evolutionary process stayed on track. He knew of Satan's plotting and waited to see what he could accomplish.

Toward the end of the Ordovician Period, Satan saw that the stage was set for the beginning of an ice age. He decided to accelerate and increase this effect. With the help of his fellow fallen angels, he caused a star that was about ten thousand light years away from our solar system to become a supernova earlier than it would have. The resulting burst of gamma radiation bombarded the earth's atmosphere and wiped out about 40 percent of the protective ozone layer. Gamma rays also induced the formation of nitrogen dioxide, a brown gas that blocks sunlight. At the end of the Ordovician, most of the landmasses were at the southern pole where large ice sheets formed. The tropical areas also cooled, causing climatic changes, resulting in cool summers, so the organisms that needed warm temperatures to survive had nowhere to go.

Over 60 percent of invertebrates and 25 percent of all species became extinct. The extinction event, called the Ordovician event, was second in severity, the severest being the Permian event. No definite cause has been found for the Ordovician event, though the possible occurrence of a gamma radiation event has been postulated. Elohim looked with sadness at the sudden loss of so many of His creatures. He tried to communicate with them and decrease their panic and helped them to find food. If the loss of these simple creatures was so upsetting, how would He deal with the loss of humans when they came into existence? He remembered the Ediacaran fauna and checked on them. Not one was alive. The cold spell had killed them all. He was saddened and relieved at the same time. He decided He would weigh things carefully every time He created a new creature. *At least they will be remembered*, He thought. In a few million years the sand would be reduced to sandstone and the imprints they left would be etched into stone. They would definitely be discovered by humans of the day. Many of the Ediacaran creatures were more complex than those that came later. This was because Elohim spent a few thousand years crafting them and enjoying their company. Now, allowing the natural processes to take their course, He watched as Satan's machinations caused several different environmental-changing events, leading to smaller extinctions. Relieved that the Ediacaran creatures were no longer a problem, Elohim began making plans for how He would create humans.

Elohim began to plan for His eventual design of man. He had command of all the laws of physics and biology and could create out of His substance anything He imagined. He preferred, however, to the lay the groundwork for a body plan for man and then add features that would make him far above all other species in both morphological features and cerebral ones. Since man would have the ability to think about and plan his life to a far greater extent than any other creature, he would need the appropriate areas in the brain to perform these functions. Unlike other species, man would eventually need to learn a series of languages. To do this, Elohim would need to add anatomic changes to his larynx. Elohim decided that He would add speech and language toward the end of His creation of man when there would be several human species living concurrently. At that time, He would give all of them the gift of speech and language.

Despite the setbacks, during the next period, called the Silurian, life proliferated again in the enlarged warm oceans. The Silurian lasted over thirty million years, from 443 mya to 416 mya. During the cold spell brought on by Satan, areas not affected by excessive cold sheltered cold-blooded species and when the climate changed, they spread to other areas. The constant movement of landmasses and creation of new oceans played a part in the dissemination of surviving species. Elohim was pleased with the interplay of climatic and tectonic events influencing the course of His created evolution. By slightly altering the timing of naturally occurring events He could speed up or slow down the evolution of a particular family of organisms and encourage the evolution of an ancestor that would produce the next step toward the creation of Homo sapiens. Toward the end of the Silurian (417 mya), Elohim introduced fish with jaws and notochords. This laid the basis for the evolution of the branch called *vertebrates* from which mammals would arise.

The Devonian Period followed and lasted from 417 mya to 345 mya. It was a time during which Elohim concentrated on the evolution of fish. He had initially introduced bony fish and cartilaginous fish. The cartilaginous fish had a skeleton made of cartilage. They are represented by sharks, skates, and rays. The bony fish are represented by various species of lobe-finned fish. These were fish which had fins that looked like stubby feet. He then introduced ray-finned fish, which were very similar to the fish of today. A strange fact: Ray-finned fish evolved in the Devonian and continued to have some new variants for a few million years. Then, as planned by Elohim, He reduced speciation, yet they are what we have in the oceans today.

For the most part, evolutionary changes inched along since Elohim had reset the mutation rate to its natural slow pace. No extinction event had occurred since the Ordovician event and the surge of speciation after that brought life back to normal. The random interventions of Satan to interrupt this process resulted in waste of life, but did not change the trajectory of Elohim's design. He did, however, insert into the evolutionary process certain limitations. To avoid the earth being bombarded with an excess of creatures, as new species evolved over time, He built in limiting factors.

One of these was the creation of branching blind alleys to evolution. These blind alleys varied between having only a few branches followed by a block to further differentiation, to branches that existed unchanged for thousands of years. As a result, most major genera stopped producing new species after a limited number of generations. Insects, for instance, reached their peak of evolution and stopped differentiating early in the Cenozoic Era about fifty-five million years ago and mammals stopped producing new genera much later in the early Pliocene, about three to four million years ago. Elohim was closely involved with the process that surveyed the evolution of mammals and often made the decision to freeze a particular line.

Chapter Nine

Devonian Period:

From Chordates to Early Amniotes

345 mya–290 mya

By the end of the Ordovician and into the Silurian, Elohim encouraged the proliferation of marine chordates and more metabolically sophisticated species of plants. The period following the Silurian, which we call the Devonian, started 345 mya and extended to 290 mya. During this period, the proliferation of the fish species continued. It was the age when a group of fish called Sarcopterygians flourished. They had bony plates covering them instead of scales. Elohim selected them as His choice of fish to turn into the first stage of four-legged creatures that would transition to land. A version of the Sarcopterygians, called coelacanths, had air bladders inside the chest cavity that could inflate or empty by pharyngeal movements. Elohim saw this as a good beginning for lungs that would enable air breathing. He planned to produce a genus that would develop four legs and would eventually be able to breathe air with the aid of lungs. They would begin the process of populating the land with animals. But that was still in the future.

By four hundred ten million years ago, the proliferation of bony fish and the migration of plants onto land were the two big advances of the time. Surprisingly, the migration of plants appeared to occur from fresh water sources rather than from the oceans. The initial colonizers were algae whose spores traveled from one freshwater pool to another and then migrated onto the adjacent land surface. Plants also proliferated in the oceans in the form of extensive algal reefs. Species of fish multiplied and grew in complexity. Elohim decided that, for now, He would give the fish some organs that would be useful when the time came to form vertebrates. The fish still had a vestigial vertebral column that it acquired when He first invented chordates, so He wanted to establish a proper vertebral column. He selected a coelacanth fish and carefully modified it. He introduced mobility into the neck area so it could look from side to side. He also fused the sacral vertebrae, which prevented its back end from sagging when the newly formed tetrapod tried to walk on all four legs. He fine-tuned the joint articulations between the legs and arms and between the pelvic and pectoral girdles. The airbladder in its chest cavity would enlarge as it made its transition to land. Elohim decided to leave this to happen naturally. The gills would become atrophic once the fish became acclimatized to using its lungs on land.

By the end of the Devonian, forty-six million years later, the first tetrapods (four-legged creatures) left the water for the land. They continued to lay eggs in the water because if laid on land they would dry out or be eaten by predators. This group of tetrapods remained dependent on a watery environment. They evolved into amphibians, and the earliest species were called labyrinthodonts, which, over time, gave rise to a wide variety of amphibian families. Amphibians still had gills. They laid their eggs in the water, covered in a gelatinous material, to protect them from predators and to keep them moist. In the Devonian, the earliest amphibians mostly lived in water with short trips to land. They lived in the ocean, in rivers, and small lakes. Elohim decided to let the amphibian line reproduce on its own. He did not need them to establish any special features that would contribute to His plan for mammals.

Before He started on the precursors of mammals, Elohim decided to create the amniotic egg. It would be an essential component of the reproductive cycles of both reptiles and mammals once they were established on land. In reptiles, the amniotic egg is enclosed in a hard shell and hatched outside the body. In mammals, it is generally generated within a special reproductive system. He picked a female tetrapod that was about to produce eggs and changed the process of the egg forming. Tetrapods at the time produced fish roe-like eggs which were released into the water. Elohim instead introduced the production of single eggs that would develop externally.

The new egg had a hard shell which was permeable to gases and provided protection from predators. The chorion was the next layer. It formed a sac around the embryo and was filled with fluid which provided a buffer to protect the embryo against injury when it moved. There was a space between the shell and the chorion that was filled with thin albumin to provide the embryo with protein for nutrition. The yolk floated within the chorion and was surrounded by thick albumin. It was held in place in the middle of the albumin by a chalaza. This is a stringy, stretchy, tubular structure that is attached to the shell membrane and to the chorionic membrane. This can be seen when an egg is cracked for breakfast. It stabilizes the yolk and conveys nutrients to the embryo. Inside the cavity enclosed by the chorion is the embryo and two separate sacs. One is the allantois. It is responsible for getting rid of waste. In mammals, there is another structure called the vitelline. It is initially the topmost layer of the early embryo and is involved in the process of binding the ova to sperm and the process of incorporating the chorion and yolk sac into the embryogenic process. The vitelline is traversed by blood vessels from the uterus and brings gases needed for respiration.

Not to keep pressing this point, but it comes up every time rapid evolutionary change is experienced. One can see from the process that Elohim established the formation of an egg, which was a continuously generated progression involving the production of hundreds of needed enzymes and catalysts. The only explanation for it occurring so fast is that Elohim fashioned the egg and started the reproduction process in a few animals, mostly reptiles and amphibians. It is hard to imagine that this biological system could have evolved one protein at a time through natural selection alone. Even a single gene variant took at least hundreds of years to enter and establish a change in anatomy or physiology. Above is an extremely simplified account of the entire process that eventually led to the reproduction of a new member of a species.

Thousands of hormones, enzymes, and organic and inorganic substances are needed to accomplish even the maturation of a reptilian egg to an embryo. As embryos become more complicated, even more control of homeostasis is needed to produce a perfect replica. It is almost a miracle that most of the time it does so. Elohim went ahead and established the process. Once it was established, all the enzymes needed for continuity would be embedded within DNA so that the process could continue on its own.

As usual, Elohim ended this bout of creation by introducing His monitoring systems for detecting threats before they occurred. He would not take the risk of problems getting out of control and ending in systemic failure. He started by establishing one segment and then moved to the next per His usual policy. He was especially cautious about new traits being absorbed into the vital reproduction system and left to insert itself and make changes. Variation could come at a cost that was irreversible. It could be responsible for the extinction of small groups of species. He assessed the egg-generating process as having a high burden of risk from adding a new gene. Therefore, He placed it on the high constraint list and hooked it up to His dark energy. He gave instructions that He was to be notified before allowing a new trait that involved the embryology of an egg. Elohim completed His invention of the amniotic eggs and was satisfied. He was one step closer to the creation of man. In a few hundred thousand years, the new eggs became a feature of all reptiles and tetrapods that reproduced using external shelled eggs. Amniotic eggs evolved in the late Carboniferous, about three hundred forty million years ago. An interesting observation: The earliest fossilized eggs that have been found are from the Triassic, about two hundred million years ago. Paleontologists feel this must have something to do with the preservation of earlier eggs, either because of the fossilization capability of the early cells or to their susceptibility to being dissolved by chemicals like carbonic acid in the environment.

Tetrapods were a successful group now living on land and increasing in numbers. They have been given the name amniotes, derived from their being the first genus to reproduce with amniotic eggs. Because of their ability to lay eggs, which had thick rubbery coverings, they rapidly evolved into land-living creatures.

Chapter Ten

Amniotes to Placentals

300 mya

Elohim paused a moment at this point to survey the reproductive systems He already had in place. Long before starting on creatures with amniotic eggs, He had introduced the viviparous process in other lesser species. Viviparity is the process where, after an egg is formed in the reproductive system, it remains in the mother and is provided with a vascular bed through which it is fed protein-rich liquid products. At the end of gestation, the birth of a mature infant is orchestrated by hormones produced by the mother. It is guided out of a birth tract to the appropriate environment.

Elohim thought of producing such a reproduction process in mammals. He noted multiple attempts at establishing viviparity in coelacanth fishes and several amphibian families. In addition, similar methods of reproduction appeared in plants. After pollination, followed by seed formation, seed dropping did not occur. Instead, new plants sprouted in situ on the plant and when ready to be planted as mature plants they fell off and set roots next to the parent plant. This often occurred in plants that died after the seed formed. The red mangrove and the grey mangrove are modern-day examples of this phenomenon. Insects and fish are another group in which this occurs especially in select species like spiders, cockroaches, and sharks. He decided to pattern a similar process using amniotic eggs which are fertilized embryos, with a vascular placental organ. Insemination would be accomplished internally or by interfacing with a female cloaca and result in the formation of an embryo. Similar scenarios had already occurred in female insects and spiders.

The first amniotes were unusually large. Elohim expected this. The evenly provided nutrition from the amniotic eggs gave the embryos a jumpstart in life. In general, larger eggs yield larger progeny in nature. In addition, more efficient removal of waste products provided healthy progeny. Elohim decided to introduce a non-mammalian line of animals that would complement the variety produced by the other two groups of amphibians and mammals. He imagined the general physical appearance of the various families to be included and created prototypes of them. He introduced various genotypes into the existing Tetrapoda clade and in a few thousand years several groups evolved. They superficially resembled lizards and crocodiles of today. They were amniotes and laid amniotic eggs with a leathery external capsule on land. Elohim initiated and formulated a name for His new group of amniotic tetrapods, called reptiles.

Elohim next tackled the much more difficult process of producing His chosen family, which are known as mammals. The initial members of this group superficially looked like amniotic reptiles. They had, however, a totally different purpose. This was the group that would give rise to Homo sapiens. They were given the scientific name of synapsids and would eventually become mammals. Elohim took stock of where He was introducing mammals. His long-term plan for them was to provide a source of high-energy food for man, initially by hunting and later using animal husbandry. In addition to this, He hoped that man would take care of them and make sure that their needs were cared for. He hoped that man would look on them as companions in the beautiful world He would craft for them. He planned to create species that lived in every environment on the earth. Although most of them had all the mammalian traits, they would be varied in appearance and habits.

Designing and creating numerous complicated animals in a few million years would be quite a challenge and He looked forward to it. He did not expect much help from the evolutionary process and introduced gene exchange by establishing the process of meiosis. He wanted a baseline mechanism in place that would ensure beneficial advancements to all life forms over time. It would be the main process to bring about the development of new traits in populations. It had worked for Him in the Cambrian Era. The traits would be produced within families and in time would spread into communities by meiosis. If a harmful trait was allowed, it would also be spread the same way. That was why He had all His defensive procedures. Elohim noted that He would not yet be able to achieve the full beneficial effects of meiosis because the full spectrum of sexual reproduction was not established. It would need the introduction of internal anatomical structures to accomplish this. Meiosis would be available through the amniotic eggs so other mammals would have access to at least the advantages of meiosis.

Elohim knew that creating mammals was not a result that could occur with random mutation, at least not for billions of years, if ever. It required careful observation with occasional intervention, a creative process He knew and enjoyed. There was no process more enjoyable than creating perfection and He was looking forward to this with almost as much anticipation as He was to the creation of man. Looking forward in time, He saw there was potential for a major extinction event in the future.

Two hundred million years ago, a newly evolving clade of animals might well be eliminated when this occurred. He resigned Himself to not starting any new processes of creation to pause and wait for the event to occur. As He noted before, a large asteroid was in route to collide with earth in about two hundred million years. He knew that the consequences to life on earth would be dire. Instead of creating mammals, He would spend time fine-tuning each mammalian feature as it appeared and let it affect a significant proportion of stem-mammals before starting work on another. He started with features He needed mammals to have, like endothermy, which is the ability to use metabolic pathways to keep a steady body temperature. The most important process would be the process of sexual reproduction by insemination in placental mammals. The process included internal insemination, retention of the embryo in a uterus, and a gestational period while the embryo matured. This would not occur until the Miocene, when placental mammals would emerge. In the meantime, amniotic eggs would have to do. Once placental mammals evolved, He would check to be sure that all processes were working the way they were designed.

A placenta provides nutrition, oxygen, and a vascular system to bring nutritional products in and take soluble waste products away. The main addition to the amniotic egg would be a placental structure that establishes an attachment to the internal surface of the uterine muscle called the endometrium. A new vascular system of arteries, veins, and capillaries coming from the maternal arterial circulation would establish itself in the uterine wall. The chorion, in the region of the placenta, formed fingerlike structures called villi that carry the fetal capillaries into spaces in the placenta that are bathed with maternal blood. Oxygen diffuses from these spaces through a single layer of cells covering the villi. From there it enters the fetal capillaries where red cells absorb it and are carried to the fetal tissue. In addition, needed chemicals like glucose are diffused into the fetal circulation and are delivered to where the fetus needs them. Nutrition, oxygen, and warmth are all delivered to the fetus through maternal blood flowing around chorionic villi. The villi are formed by a single cell membrane and gases diffuse across it into the maternal blood and vice versa. Carbon dioxide, waste products of metabolism, and excess water are all carried away from the embryo and discarded into the allantois. As blood evolved in complexity it even brought antibodies to fight infectious agents and other nutrients that were needed for special purposes.

Elohim started the process by changing the embryology of a female mammal to have the capability to evolve a uterus and all the other structures involved in the reproduction of a baby. It would have to start with an embryo. He would have to chart the anatomy and insert into the current anatomy a mammalian species without a placenta and uterus. The current structures were optimized for egg production, and egg release could be adapted to provide the fallopian tubes and ovaries. He had to insert a uterus and a vagina. He also had to introduce new brain cells to produce the hormones that gave signals to the various structures involved to prepare for and initiate reproduction.

Once He had a detailed plan in His head, He rearranged matter and added genes which would establish the new process in every structural feature to produce the correct anatomy and physiology associated with the process of reproduction. It is impossible for us to accomplish, but to Him it was effortless. The planning and accomplishing of His ultimate goal was His favorite activity. He also initiated areas in the brain that produced the hormones that changed the female from a dormant state to a fertile child-bearer. Elohim thought about introducing this complex system in a group of mammals to make them placental mammals. It was going to be a complex but fascinating task.

Elohim considered the intricate workings of all He had to do to accomplish His dreams. His most urgent purpose was creating a species of human that would closely endorse and practice the moral code He would teach him. At some point He had to stop creating sporadically and start a smooth process with which He would be occupied creating, modifying, and improving in a variety of creatures. It would involve so many different aspects of matter. Perhaps He should use His abilities in maneuvering matter to fast track this process. He decided to attempt this alternate method of creating. It was a method He had imagined and prepared for, a kind of intangible creating out of nothing whatever creature He imagined.

He focused on a pair of mammals and chose opossums. He would start with an image of what they would look like. A male and a female. Then He would look at the mammals that had recently appeared in His epoch and choose a pair that was close to His image. Having a three-dimensional construct of the anatomy, He would convert it to a body plan. He would then do the same for the male. Using His ability to rearrange matter at all levels, He would then reproduce the body plans. He went through the process and had no difficulty at all. There before Him were two opossums typical of those on extant earth. Elohim thought about whether He wanted to continue to produce living things using the laborious body plan design or if would He create a new method. At the end of the laborious process, He felt fulfilled and closer to His creations and decided He would continue the process that He had been doing since the beginning but find new ways to process some of the steps so they would be done automatically.

He considered what else He would have to introduce to make the process of birthing safe for His humans. As He looked at potential body plans and ways to increase the size of the offspring, it became apparent to Him that the structure of the pelvic outlet would be a limiting factor. Rather than change the body plan He had envisaged for man and woman, He decided to create a hormonal signal system that would initiate the birth process while the infant was still immature. The baby would be born after a gestational period of thirty-eight weeks and be dependent on its parents for many years as it matured. Most other mammalian species, including primates, would bear infants whose heads were small enough to exit the natural passage designed for them.

Man, with his large head, would be problematic. He decided to increase the stretch capability of the female reproductive structures as well as to allow for the average size of an infant head at thirty-four weeks of gestation to pass through the pelvic outlet. In addition to decreasing gestational time, Elohim decided to slow down a human baby's rate of maturation. The baby would remain juvenile for much longer than the average ape. This would allow time for the brain to continue to grow after birth. Elohim would use the childhood period to prepare the child for a much more complicated adult existence than all other species.

He introduced all these new features which would enable safe and effective sexual reproduction in a basic unspecialized mammalian body plan. He would use this as the basis for specific body plans for each mammal, introducing special features if needed to fit the elements of the basic plan. All the morphologic features that enabled giving birth, and taking care of the infant after birth, would be incorporated into each species plan to fit with His specific plan. One important element was the incorporation of the female's lactating breasts to keep the baby nourished during its first year of life.

This state of immaturity is called neoteny and is a universal human characteristic. The baby would need the intensive involvement of a parent for the first few years. Once caught up, they would be like other primate babies, continue a gradual transition from infant to adult. Chimpanzees, for instance, are carried around clinging to their mother's chest after delivery. Later, they ride on their mother's back, holding on to her fur. In general, apes and monkeys have shorter juvenile states than humans. Humans are literally helpless at birth. They become physically independent at two to three years, but even when the motor system is increasing in independence the brain will be growing and learning well into their teens. Maturation to adult form would take place at about eighteen years.

Chapter Eleven

The Carboniferous:

Amphibians and Early Mammals; Use of Body Plans for Mammals; Genetic Regulatory Networks (GRNs)

354 mya–290 mya

The next period, the Carboniferous, extends from 354 mya to 290 mya. Elohim continued manipulating the cell lines of amniotes and amphibians to produce more diversity. To the tetrapod amphibians, He introduced two new groups called lepospondyls and temnospondyls. Temnospondyls resembled frogs and lepospondyls resembled salamanders and snakes. These amphibian families continued through the Carboniferous Period with varying degrees of success. As they entered the Permian Period they evolved into some very large species and increased in number; a few were crocodile-like creatures. Around 279 million years ago, they were challenged by the descendants of synapsids, called therapsids. These were large ferocious creatures that lived comfortably on land. They chased the remaining amphibians into their water-based habitats, and they were unable to forage on land. Even before the Permian-Triassic extinction their numbers had fallen significantly, as they did not survive the event. Modern amphibians, called lissamphibians, are supposed to have evolved *de novo* in the late Triassic. The findings do not show a clear phylogeny connecting them with the original Carboniferous amphibians. However, most scientists do not believe they are an independently evolved group since there are similarities between temnospondyls and lissamphibians, as modern-day amphibians are called.

Extant amphibia include frogs, toads, newts, caecilians, and salamanders. The caecilians of today look like worms but have lungs like amphibians and other features that do not relate to worms. They bear some resemblance to the amphibians of the Triassic. Fossils have been found from the late Carboniferous that show them to have a variety of body plans. Many looked like the frogs and toads of today. Toward the end of the Carboniferous Period, some amphibians had a cylindrical body and appeared snake-like. Others looked like alligators. Note that the alligators that are known today would appear millions of years later.

All these ancestors of amphibians are now extinct, most of them having perished in the Permian-Triassic extinction event. Furthermore, the disconnected lissamphibians of today are not faring well. A survey was conducted and published in 2004 which showed that nearly one-third of the population is at risk of extinction. Various environmental and habitat-changing issues were identified as causing the decrease in populations. A follow up report in 2021 showed that human activities that affected the decline included the trapping of these creatures for food, medicinal usage, or as pets. The affected animals composed 17 percent of the population. Elohim is saddened by this occurrence.

Meanwhile, Elohim had begun His planning for the creation of mammals. He proceeded to influence the synapsid line that He started as the precursors of mammals. The synapsids were doing very well with the establishment of the amniotic egg. Under His guidance, many of the stem mammals hatching from eggs were beginning to acquire mammalian traits, although they still looked very reptilian. Several clades arose and one of the successful ones was called Sphenacodontia. Elohim caused many mammal-like features to appear among individual synapsids. Over time, mammalian traits were disseminated within stem mammals and new clades were defined. Elohim guided this process to ensure survival of prototypes He approved of and fine-tuned each new clade. He elected to do this in stages, creating two or three animals at a time. He would start with primates, which were scheduled to evolve in approximately eighty-five million years.

All changes He made had to be incorporated into the embryological body plan of each clade. Then Elohim would enter the genetic information that initiated the trait to be transferred into a chromosome and linked to the correct body plan. When the developmental process was initiated, it allowed genetic coding to be entered into the relevant DNA strand and produced some features typical for the organism. Once transmitted to a newly formed organism it would be perpetuated in progeny. The progeny would then continue to disseminate the new mammalian trait to other members of the clade. For His part, Elohim monitored and organized the process so that it would all flow smoothly.

As traits developed in individual animals, He made sure that the final body plan was consistent with what He intended for the species. This was especially for the brain of each species. Each mammalian trait demanded special changes in various parts of the brain, so each time a new trait was introduced Elohim had to make the needed adjustment. In fact, Elohim could already see that He would need to set some automatic gene manipulations in place to guide the creation of over six thousand individual species of mammals that He planned to create so that they would be different from each other. He settled on making the body plans suitable for more than one species. To make this possible, He scripted the genetic design for embryological development to have segments that produced morphology that was very similar. Many different species of mammals would have blocks of embryological development that were almost identical so that it appeared that they belonged to the same species. The resultant adult forms, however, would emerge exactly as they were supposed to be for each original species. Elohim patiently and lovingly crafted His mammalian prototypes. He really enjoyed this part of His creative activity.

As He systematically went through establishing the body plans, Elohim decided that He could take a shortcut in the organization of this process. He set up genetic regulatory networks (GRNs) for each body plan. These are known to scientists today and are called homeobox (Hox) genes. The main unit was built for stability and would remain stable in controlling the design and patterning of the basic body form. Individual subunits would be genetically variable and would be controlled by the Hox genes. The second set of genes was called TALE genes (three-amino-acid loop extension).

The embryological development process is controlled by an interaction between a Hox gene and its complement in the TALE family. There is a Hox-TALE complex available for each phylum of animals that is represented in the body plans. The number of phyla represented in a particular body plan would each have a Hox-TALE complex dedicated to the production of a particular body part. Since most phyla are bilaterally symmetric, two halves of the body are produced, each identical to the other working from head to toe. Hox genes direct the production of a particular symmetrical body part. The same Hox gene can direct the development of equivalent body parts in different species. For instance, if the gene is supposed to produce a leg it will do so in either an insect or a bird producing the correct anatomy at the end in the final product for each order. Hox genes accomplish this with the help of a cascade of genes downstream. These genes combine with complementary TALE genes and the complex binds to downstream genes and turns them on so that the development process can begin. Eventually, part by part, the animal whose genetic code is in the embryo comes forth as a complete embryo representing the species. Because Hox genes are so strategic to the proper development of embryos, Elohim introduced severe constraints in their being replaced by any acquired genetic traits.

Scientists have confirmed the presence of these genes by experimental tests. In one such experiment, the Hox-TALE genes from a sea anemone introduced into a fruit fly that lacked the genes, produced a normal fruit fly. The anemone gene package was able to activate several patterning genes in a frog. Finally, genes from a prokaryote, acanthamoeba castellanii, were used with TALE genes from the anemone. They did not interact in any way with the Hox-TALE genes from the sea anemone.

Since acanthamoeba is a prokaryote, scientists concluded that the Hox genes only appeared in the evolutionary chain after eukaryotes evolved.

Having installed an automatic security and management system to prevent deleterious random variations from producing death or deterioration of an organism, Elohim felt that He had to embed the genetic system with some guidelines to decide if the new trait needed encouragement and facilitation or should be prevented from developing to its full potential. The whole idea of using new traits to produce a variety had been His. Now that He had brought about a substantial number of living things with continuing diversity, He decided to put restrictions on the natural process, choosing only the best traits to develop naturally.

Hox-TALE complexes were not to be interfered with, except by Elohim. He requested that He be notified if such an insertion was planned. He primed the system with the policy that changes to systems involved in the development process should be constrained but that some loosening of this policy could be tolerated or even be regarded as innovative when used in more peripheral development areas. So, that is the guideline He set in place. He made an exception during the age of dinosaurs, when He let nature take its course. For two hundred million years, He was cautious and involved in the development of new species.

To reduce the risk from an endless development of new species, Elohim decided that He would limit the number of body plans to just thirty-five and include a genetic trigger that would prevent any attempts to add more embryological body plans. Today, there are about sixty-five hundred species of known mammals on the earth. Since the Cambrian explosion five hundred million years earlier and even to this day, no new body plan has been detected at the phylum stage or higher. Most new variations occur at the species stage.

Scientists have deliberated many theories trying to solve the puzzle of limited body plans. As research continues there are two questions scientists have not answered: "How did body plans begin?" and, "How is it that the number of body plans stayed the same for five hundred million years after their inception?" To ensure that nothing would go wrong, Elohim decided that the body plans in existence would be finalized by Him, and He would create a DNA identity for each of them that would be passed on from an animal to its progeny and be incorporated into the genetic code for perpetuation.

As millennia passed, various prototypes of mammals began to form. About two hundred ninety million years ago, in the early Permian (290 mya to 248 mya), advanced synapsids called cynodonts existed with various mammalian traits. These included middle ear bones for hearing, which had migrated from components of jaw bones; stereoscopic vision, with limbs that were placed vertically under their trunk; body hair; and a large brain. Elohim was not yet ready to direct the evolution of mammals from these early ancestors. He would wait and see how the various reptilian lines developed before turning out clades of mammals. There was at least one catastrophic extinction event ahead related to a large asteroid that was headed to collide with the earth. It was still about two hundred million years away. Elohim decided that it would be too soon to fully establish mammals. He was worried that they, being so small, would be wiped out by the upcoming meteorite strike. He felt He had enough time to produce them and be ready for man if He started the process again soon after the extinction event. They would remain in the form of small shrew-like animals until the dinosaurs that He was expecting were extinct. That would also be the time He would be evolving primates. He would give priority to primates and work at a slow pace with other mammals.

While Elohim involved Himself with mammals, to provide new animals that would provide both food and companionship for humans, many other events were occurring on the earth. Elohim worked on the progression of plants to their ultimate forms. Trees evolved with trunks and were made of woody lignin and impregnated with xylem and phloem channels for carrying water. The support provided by the woody substance enabled water to travel to greater heights and tall trees evolved with the proliferation of swampy forests. In these tall, dense forests, the extent of photosynthesis was so high that oxygen levels in the atmosphere were 30 percent instead of the 21 percent today. This produced giant arthropods. They included a scorpion that was fifty centimeters long and a dragonfly with a wingspan of seventy centimeters.

This was a time when major changes were occurring in the landmasses that would eventually form the continents. In the first half of the Carboniferous, the continents of Gondwana, Africa, South America, Antarctica, and Australia began moving together. Polar ice caps were still present. As an ice age set in and the extent of the ice increased advancing as far as thirty degrees from the equator, two landmasses were separated by a narrow sea called the Rheic Ocean. During the Carboniferous, the continental shelf of the two landmasses collided, forcing upwards the Appalachian Mountains in what would eventually be North America. By the end of the Carboniferous (345 mya to 290 mya), the eastern islands representing the future China, Malaysia, and Euramerica merged with Gondwana to form the supercontinent of Pangaea. This large landmass lay toward the southern pole and soon the southernmost regions were covered in glaciers.

Elohim noted the event and set about adjusting the flora and fauna in a suitable manner to ensure survival of species or clades that He felt needed to survive. This was a constant process for Him. With His connection to all the universe through dark energy He knew every phenomenon that was happening and could make timely changes. In the shallow seas fringing the continent of Pangaea, the proliferation of invertebrates like shellfish continued, while varieties of bony fish with notochord increased. Placoderms, or armored fish, however, became extinct. This was a period of active speciation and many of the precursors of modern fish came to be. Sharks were one of the species that increased in number and variety. Later in the Carboniferous, the center of the continent became drier as rainfall decreased due to available water being trapped in glaciers. The center, far from any ocean, demonstrated a continental climate; high temperatures in summer and extremely cold winters. The swampy forest disappeared in central parts of the continent but there were still forests in swampy areas in coastal regions.

Chapter Twelve

The Permian-Triassic Extinction:

A Series of Giant Reptiles

290 mya–248 mya

The Permian is a forty-million-year period after the Carboniferous Period. In the early Permian, many types of reptiles walked the earth. At the end of the Permian, only a few amphibian and marine reptiles remained. They were products of the early reptilian branch prior to the introduction of amniotes. Elohim turned His attention to the remaining tetrapod-related reptiles who were all amniotes. They produced and laid amniotic eggs and because all three groups looked reptilian it was difficult for paleontologists to identify to which species a particular fossil belonged. They found a feature which consistently divided the fossils into three groups. The three sub-classes are anapsids, synapsids, and diapsids. Anapsids gave rise to stem reptiles and freshwater reptiles and were small and did not pose a significant risk to mammals. Also, anapsids did not have the fenestrae which are present in the skulls of the other two groups. Fenestrae are windows in the bony skull that ligaments of powerful muscles are used for chewing tough foods.

The next group were the mammal-like reptiles called synapsids, which Elohim evolved from small animals to the very large ferocious beasts called therapsids in the late Carboniferous. They were identified by one set of fenestrae on each side of the skull behind the orbits. They continued to dominate the Permian and early Triassic. Eventually, they would evolve into mammals, as reptiles laid amniotic eggs.

The third sub-class from the amniotes are called diapsids, which gave rise to Lepidosauria, which produced Sphenacodon. It also gave rise to two populations. The first produced lizards and snakes (Squamata), while the second produced carnivorous reptilian creatures like Dimetrodon (sphenodontids). The latter became extinct at the end of the Permian. Squamata continued into the late Jurassic at the same time as the dinosaurs. They gave rise to snakes and lizards which continue as extant species today.

The taxonomy of the constituents of the third group that became dominant is rather confusing with many sub-classes and orders. The members of this group are called diapsids. They had two openings on each side of the skull lateral to their orbits. It is an important group because it forms the root of several different superorders. In the early Permian, a grouping from the diapsid branch continued over time into the Triassic when it separated into three consecutive offshoots. They produced plesiosaurs, placodonts, and ichthyosaurs. Multiple aquatic creatures also evolved from this branch and became extinct in the Cretaceous Period.

The main diapsid group continued as the Archosauria branch with an immediate split which produced the large reptiles called archosaurs in the late Permian. They soon challenged the synapsid therapsids which were no match for them. A second group gave rise to the Crocodilia order in the Cenozoic Era.

They were not done. The main archosaur stem continued and when the Permian ended it produced a branch that eventually produced flying pterosaurs in the late Jurassic. The archosaur branch ended in the mid-Triassic, creating two branches. One produced the dinosaurs and the other gave rise to a new family in the Jurassic called Aves, or birds.

After the archosaurs became extinct over two hundred million years ago, their line produced a new superorder called Dinosauria, a series of large reptilian-looking creatures with bizarre features. They were accompanied by smaller species, and were both herbivores and carnivores.

Once again, dramatic battles for food and territory became a perennial theme and the archosaurs gradually reduced in number and disappeared. Elohim was sad for them, but He had another evolutionary drama pending. He was content with the pace of progress in mammalian evolution, so He turned His attention to accelerating the process of creation in both marine and land creatures. Most marine life at that time was sessile with various species that attached themselves to the seabed. He saw that there was a need for these creatures to be mobile if they were to provide a complex nutrient cycle for larger creatures. He went ahead and increased the species of fish, diatoms, and other marine creatures that could be a food source for larger animals.

Elohim was fascinated by the flying pterosaurs that had evolved from the archosaurs. He had fond memories of flying creatures from a previous universe He had created. They developed into several sizes from small to very large. Pterosaurs had wings made from thin membranes which attached to their very long upper limbs and elongated claws, similar to bats. Most of the small and medium-range creatures could fly, as they had wings and thin membranes, providing them with good aerodynamics. They also had streamlined bodies and bones which were hollow. He decided that He had to find an equivalent airborne creature to populate the world for another hundred million years. If dinosaurs proved to be too large and too violent to last, He would need another family of creatures that would have a better disposition. Man would not last long in a dinosaur-controlled earth. He decided to allow and even assist dinosaurs to become extinct at the end of the Cretaceous. That was the period when a large asteroid would impact the earth, and the dinosaur population would be severely reduced. The absence of the dinosaurs would again take Elohim's evolution to a gentle steady pace.

As the Cretaceous ended, it experienced scenes of massive destruction with disruption of the kind of order He liked. Despite this, He enjoyed the amazing dinosaur drama and even became fond of them. He loved the fact that they appeared to be violent mostly when in search of food or if there was a threat to their eggs or family. Not all were carnivorous. The herbivores were both small and very large in size. Some small dinosaurs were the fierce predators that He had been waiting for. Unfortunately, He was not sure that this would balance out. There was more destruction than growth and the predator-versus-carnivore ratio was on the side of the large and destructive carnivores.

The massive destruction by the dinosaurs had already introduced so much instability in the evolutionary process that Elohim would have to spend a few hundred thousand years re-establishing a slower, more controllable pace in which He could carefully work to produce mammals as an alternative group of animals so they would co-exist with man. He decided to follow the evolution of a new dinosaur species. He would try to balance their size by selecting a few large fossils to leave for man to find. Looking at the various groups of animals from the Carboniferous to the end of the Cretaceous, scientists discovered that the Carboniferous had eight major lines of evolution. They all derived from the creation of amniotes. Synapsids slowly developed, but once they reached the Triassic as therapsids, no further branching occurred. However, by the Cenozoic, a full slate of mammals evolved.

The branching of new species can be quite confusing especially when following the progress in terms of two timelines—that of the evolving earth and the knowledge of the earth at its current era. Anapsids also derived from the Carboniferous to the Cenozoic with no branching and they became stem reptiles and freshwater reptiles.

Next, the lepidosaurs, the first reptilian amniotes, persisted as a line of evolution and made it to the Cenozoic as two separate lines. The first group made its way in time without major branching to become part of the Cenozoic as two additional evolutionary lines. One line produced snakes and the other line produced lizards. A separate line from the amniote origins made it to the Cenozoic intact and was responsible for producing Sphenodontia, which are small marine reptiles.

Finally, there is the branch of diapsids. They separated from amniotes and began evolving aggressively while still in the period of the Carboniferous/Permian. They split into two branches in the Permian. One branch, the original diapsids, began separating into three units soon after entering the Jurassic Era. These three branches produced placodonts, plesiosaurs, and ichthyosaurs.

The main branch continued with only archosaurs until the end of the Permian when it broke off into two separate creatures and continued into the Jurassic Era and then into the Cenozoic, where it gave rise to crocodiles and other reptilian creatures.

Dividing from the same location as the previous offshoot, a single branch journeyed into the early Cretaceous and produced pterosaurs. These are flying reptiles also called pterodactyls in lay terms. The main branch, which had divided into eight branches already, split into two branches that produced two orders of dinosaurs, Ornithischians and Saurischians. The main branch persisted in time until the middle of the Jurassic Era when it produced birds.

Birds are classified as dinosaurs because they were an offshoot of the dinosaur branch. Dinosaurs diversified during the Cretaceous and were as large as the archosaurs. They soon became the dominant actors in the Jurassic and Cretaceous Periods. The lepidosaurs continued through all the periods ending with the production of crocodiles and alligators in the

Cenozoic. Dinosaurs occupied the earth for about two hundred million years, starting in the early Triassic and ending in the late Cretaceous (250 mya to 60 mya).

As giant reptiles came and went, a small group of early mammals survived by keeping themselves out of the path of the giants. They were small rodent-like creatures called cynodonts, meaning dog-toothed reptiles. They would become the direct ancestors of true mammals. With the cynodonts, there were also many stem mammals that Elohim had evolved and moved into locations safe from the dinosaurs. He had begun miniaturizing them so that after the Permian/Triassic extinction event they could remain hidden from their giant kin. They were now the size of small shrews. Elohim had already begun taking preventive measures to shelter stem mammals from the large lizards. Gradually, He miniaturized the groups of stem mammals that evolved in the Carboniferous and coerced them into nocturnally browsing groups that lived in burrows. He enabled a social element to their activities, feeling that a group defense would be more effective in dealing with the dinosaurs. He would take no chances on these mammals being wiped out. His human species was to have biological features found in the mammal family.

The Permian/Triassic extinction turned out to be the largest extinction event in geologic time since life began. A study of various isotopes contained in rocks aged around two hundred fifty million years ago reveal that there were multiple unrelated natural events occurring consecutively that contributed to the large percentage of organisms that were obliterated. These effects were especially prominent in the oceans. The main event was most likely the eruptions of lava and hot gases from what are called today the Siberian Traps. The source of the emissions was a large plume of molten material from the earth's core which made its way through the lower mantle into the upper mantle and then spilled over. Multiple small plumes originating in the large spillover area sent plumes of lava, gases, and smoke into the earth's atmosphere. This was an intermittent process that continued for years as the plume was constantly replenished from the core. The result of the eruptions was a massive increase in the carbon dioxide content of both air and water over time.

Examination of rocks in China showed that 286 out of 329 marine invertebrates were eliminated. On average, over 90 percent of marine creatures died since their simple metabolisms could not tolerate the high carbon dioxide levels of the ocean. Most of these organisms were creatures with calcium carbonate skeletons which needed stable CO2 to make the matrix of the skeleton. The acidity of the ocean from the large amount of carbon dioxide dissolved in it (carbonic acid) caused the shell to deteriorate. One hundred percent of the acanthodians, blastoids, and trilobites (sea creatures which were already in decline) became extinct, mainly due to carbon dioxide toxicity. This did not occur in all ocean areas so pockets of many species that were mostly wiped out survived and multiplied when the environment became favorable. Land animals were more resistant to the high carbon dioxide levels and more of them survived, but invertebrates and insects were hit hard. Remember, this was the era of giant insects. Eight or nine insect orders became extinct. This gradual extinction of most of the existing species suited Elohim's purposes. Many redundant species, which were offshoots of the pre-mammalian synapsids, were eliminated. Now, He could restart the population of the oceans with more mobile creatures to replace the numerous sessile ones that had died.

Although plants were not directly killed by the toxic atmosphere, the changes in the ecosystems, especially the disappearance of the forests, led to widespread patchy loss of flora. Gymnosperms and seed ferns went into decline and were replaced by lycophytes. Herbivores that depended on them as a food source were affected by this decrease. Gymnosperm forests would not reappear for another five million years.

Terrestrial vertebrates were also less affected than marine creatures, but the cost was still high. Sixty-five percent of the amphibian, sauropsid, and therapsid families became extinct. All anapsid reptiles died off during the extinction. These are the groups that would have given rise to lizards, snakes, crocodilians, and turtles that came from the original tetrapod amniotes. Considering they had all been created in the Permian, this was a premature loss. The herbivores suffered more losses than other families and were replaced by diapsid reptiles. Careful studies of all possible events that impacted the Permian extinction initiate more hypotheses than well-understood mechanisms. That an acute event occurred two hundred fifty million years ago is certain, based on the data on marine and land organism losses. Marine creatures fell to the lowest levels in history. Since CO2 toxicity explains this phenomenon, the assumption was that it happened when the Siberian Traps were actively venting. In addition, several other events and mechanisms may have occurred that contributed to the huge losses. We do know that it took thirty million years to recover from this setback.

Elohim surveyed the conditions at the end of the Permian and was satisfied that the damage was not permanent. His mammals survived by staying in their burrows. Soon He could start His work on mammals. The insects would recover as they were indestructible. The plants were destined for constant changes in eco systems and had already learned how to adapt to them. They may come back as different flora, but green plants would inhabit the earth again very quickly, armed with their metabolic processes to clear the acidity of the oceans and the murkiness in the atmosphere. This was the order that He had already introduced into His universe. The amphibians were a sad loss, and Elohim would create them again.

Chapter Thirteen

The Jurassic Period and the Evolution of Birds

208 mya–146 mya

It took about ten million years for the earth to recover from the Permian extinction. As things settled down, the dinosaur order proliferated and enormous, bizarre creatures occupied the earth. Elohim found them to be fascinating. They were the dominant occupants of the period and survived on a variety of prey but also laid eggs, hatched them, and took care of their young. They were barely aware of His presence at first but as time passed and He communicated with them, they became accustomed to Him and welcomed His contact. Elohim tried to balance the violence engendered by their enormous size by consoling the smaller creatures and at times protecting them. He was especially careful to continue to keep His mammals hidden.

As the Jurassic progressed, Elohim thought about the impending disaster that was headed to the earth. He decided to stick to His decision to let evolution take its natural course. He would spend His time on a new project. He had noted that there were several creatures that could glide or fly. They were pterosaurs. They had wings made of a thin layer of membranes as well as re-enforcing muscle fibers. They could launch themselves from the ground to take flight. They were not exactly what He was planning, but in another universe, He remembered flying creatures of various sizes and exotic colors. They ate fruits, insects, and seeds. The most beautiful feature about them was their singing. He would create a feature like that for His humans. He looked at the existing dinosaurs and saw that in the previous hundred million years that many of them had randomly evolved feathers. The feathers were of various colors and shapes and often found on the body. The genetic variations that produced them had not yet given them an edge to selection because the only advantage they provided was warmth, which could have positive or negative effects depending on the climate. Regardless, this gave Elohim several possibilities for Him to conceive a new order called Aves (birds). If He adapted the existing species of dinosaurs in a sequential manner so the required elements to create a bird that flew were present in the final product, He could do it in a quarter of the time it would take to create a new phylum from scratch.

A big advantage would be that all the dinosaurs He used would already be adapted to the environment in which they lived. Another advantage was that, because of this, the new phylum of creatures that would result would not need a fresh body plan. Starting from a dinosaur blueprint and creating a bird off-shoot meant He would simply be producing flying dinosaurs in stages. The embryology would be very similar, so He decided to proceed with His idea.

He found a group of small dinosaurs that had a skeletal structure that could be adapted. The dinosaurs were a high-energy group of small to medium-sized carnivores called therapods and had hollow bones and claws on both upper and lower limbs. He chose one that was a medium-sized creature. From His previous experience, the flying creatures had ranged from tiny nectar drinking birds to those who stood five feet high and were flightless. A small size would be a major factor to enable flying. He decided that He would give them more intelligence than most animals of that size, making them a communal type of animal that lived in flocks and talked to each other in musical sounds. They would evolve in varied colors and would be found in numerous environments, each area having its own unique population. They would be herbivores and carnivores, and like dinosaurs, would reproduce by laying and hatching eggs.

Elohim started with a sub-group called Ornithischian dinosaurs. He chose a euro raptor and made some changes in the development of the embryo. The progeny of these dinosaurs had a decrease in length of the fourth and fifth digits in the feet. Adapting one or two features at a time and crossbreeding the animals so that the changes would accumulate in each successive species, Elohim crafted a facsimile of a bird. It was not a flying bird as is known today, but was similar to an ostrich. He did, however, establish all the morphological changes needed for flight and increased metabolic capabilities for the act of flying.

Along the way, the bird's four-digit hand was formed, and the bones became hollow. The next advance used tyrannosauroids, which contributed a three-digit hand and a pubis to the model. Elohim looked for feathered dinosaurs to add the next feature and He found them in an oviraptorosaur, who had hollow cylindrical and tufted feathers. He was pleased with His product so far. With all its bird-like changes, there were only three more steps to a reasonable facsimile of the first bird. The arms became wings, and the lower limbs became shorter than the arms. The creature still had teeth and a blunt tail. A bird evolved from a sequence when Elohim contrived the genetic changes that contained the embryological instructions written by Him to produce a totally new type of living creature. This did not happen overnight. The crossbreeding process used to accumulate the changes into one bird took chunks of time.

But could the bird fly? One step at a time Elohim created a bird using a difficult technique of crossbreeding using suitable dinosaurs to accumulate the traits a bird needs to fly. He used crossbreeding because He enjoyed creating in this manner, mimicking the method as it happened in real life. The difference was that without His skills and knowledge it would take hundreds of millions of years to evolve, instead of the approximate fifteen million years it actually took. Even though it took about fifteen million years, it is amazing that the process occurred at all. It happened in a stepwise fashion going from the most-needed features to the finishing touches. Elohim was pleased with His initial avian representatives. Fossil studies have suggested that the first bird was an archeopteryx, which was derived from its own separate phylogenetic path. It took Elohim ten million years to produce the first birds, and they were still very large. Elohim decided to start miniaturizing them slowly. He gave them the physical features that could initiate simple flights like flying from a tree. They could not, however, take off from the ground. Some of the creatures He produced needed the instinct to fly, unlike most birds that exist today. In fact, baby birds today are able to launch themselves and fly without too much coaching. Once the prehistoric birds obtained the instinct to fly, it would eventually be transmitted genetically from generation to generation. He introduced the ability to fly into a couple of the newly evolved birds; in others there was the urge to fly, but they were ignorant of how to do it. At first, they started by climbing trees and jumping from the branches and later figured out that by running they could *take off* and fly. Elohim, for His part, tried to encourage the evolution of smaller species in whom flight dynamics would need much less energy for takeoff.

Elohim continued to produce birds with many different colors of plumage. He also induced clades that were carnivores and others that were omnivores. Some could live on water and others loved forested environments. By the time they became modern birds, they were one of the orders that had a large number of species. It did not take long for the smaller birds to start flying by taking off from trees which they were able to climb. It would take longer for the rest to learn to fly, chiefly because of their size. Reducing the size of the birds was a gradual process.

Paleontologists have long been amazed at how fast birds evolved from dinosaurs. Some have called it *miraculous* that a completely new order could be achieved by natural selection in such a short time of ten to fifteen million years. They have suggested mechanisms for the phenomenon. One developmental scientist has studied the evolution in birds and has done experimental research on the genetics of its evolution. He feels that the sudden decrease in size from dinosaurs to birds was not an expected phenomenon, as most other species were growing larger. He suggests that the mechanism of the rapid drop in size is due to a process called pedomorphosis in which the genetic process halts growth and development at a juvenile state so that a smaller adult results. When development is halted (like neoteny), size is also arrested at that age. Biologists have seen this pattern in other evolutionary paths. The scientist also describes an alternative view of bird evolution that could explain the rapidity of how they evolved.

Until the 1990s, there was not a huge number of fossils to accurately evaluate the timing of evolutionary changes in birds. In 1990, a large cache of bird fossils was found in China in Liaoning, spanning a long period of time from 130 mya to 121 mya. The Jehol collection in western Liaoning produced beautiful bird fossils. They were embedded between white tuffaceous sandstone and lacustrine shale, both of which were uniquely optimal bedrock material for preserving feathers. They are extremely well preserved and were found with a great variety of species. The type of rock they were embedded in showed the architecture of various fossils in fine detail. Other delicate structures were also well preserved. Though details are not available, the trajectory of birds from their initial formation from a variety of dinosaurs reveals familiar dinosaurian features. This occurred over 130 mya to 121 mya, and eventually culminated in the appearance of extant birds.

However, the fact that these fossils changed over a long period of time did not change the timing and existence of the diminishing body size over a short period of time. That period in the Cretaceous was one where most creatures were increasing in size to match the dinosaurs. Birds (and mammals) were the only groups that were diminishing in size. Plus, the gap that was found in the Cretaceous for fossils of birds also exists in the Liaoning collection. All these scientific suggestions explain that a decrease in size is possible, though how it happened is up for debate. Elohim may have included pedomorphosis as a technique for miniaturization. Or He may have directly changed things by manipulating the embryologic development of each subject through gene manipulation. No random process can substitute the complexity and infinite possibilities that exist with gene manipulation in the hands of Elohim.

Elohim had knowledge of how all systems work and the ability to manipulate them. Because of His knowledge, when He worked on advancing an evolutionary end point, it happened much faster and produced better results. When it happened with several traits within a short period of time, it could be given the name *punctuated equilibrium,* coined by Stephen J. Gould and Niles Eldredge. It is hard to explain a rapid evolution with a new well-adjusted species that occurs sporadically as a natural phenomenon. Even the term *punctuated equilibrium* is only a description of what happens. It does not explain why or how it occurred.

Elohim's final touch for birds was introducing the beak. It gave them the ability to do many actions that they could not do using only their feet. They came in a variety of shapes, and most had a functional intent. They would use their beaks for preening, nest building, feeding babies, feeding themselves, catching fish, and catching other prey in flight or on the ground and in water. They also used them in self-defense. It was Elohim's ultimate gift to them, even though they were probably not aware of Him other than knowing that a loving presence was controlling their destiny.

The direct involvement of Elohim in the creation of birds is explainable, as they have characteristics like no other family of animal. He had been so impressed with them that He was determined to make sure that man, when he came to be, would have the benefit of their presence. Modern-day humans prove that He was right. First, they are the only animals that fly, other than bats and a few flying mammals. Another unique trait is the incredible variation in their plumage and appearance. Thirdly, they are the only animal (other than man, whales, and dolphins) that sing when they talk to each other. Humans cannot fly, but they do create activities where they can simulate flight with assistive devices and feel as though they are flying. Think of pilots, bungee jumpers, sky divers, and skyscraper parachute jumpers, many of whom are addicted to their sport. Most people love birds and lovingly care for and feed them. Humans also love to sing and often their singing is directed to Elohim who takes joy in listening.

Elohim loves His birds and is always ready to comfort them when they are alone or frightened. Second to humans, they have a special place in His heart. He created them to give joy to man. Ironically, it is of interest that the only other family of creatures that He followed (other than the Ediacaran fauna) with intense enjoyment were the dinosaurs, from which the birds were imagined and evolved by Him.

While He was producing mammalian ancestors and preparing them for survival, Elohim also turned His attention to what He would do until the asteroid struck. He could allow any bizarre creature to evolve no matter what chaos it caused, knowing that the asteroid would wipe out most of them once it struck. He could clearly see in His vision that in the aftermath of the collision, large animals would have a difficult time surviving because of their greater nutritional needs. He became rather fond of the large lizards and thought how wonderful they would be for man when he was fully created. Looking back at the history of His universe and seeing the huge beasts left behind as fossils in rocky cliffs in every continent, He decided to let nature take its course with the big reptiles.

In the middle of the Triassic there were many diapsid archosaurs and dinosaurs that were beginning to take shape. Elohim's quick look at the dinosaurs in their evolution disclosed an interesting pattern. The predominant lizard was still part of the archosaur group which arose from the diapsid branch and was responsible for being the ancestor for the dinosaurs. They rapidly increased in number and in size. By the end of the Triassic, the archosaurs became extinct, and dinosaurs became the prominent species of the Jurassic and Cretaceous. The taxonomy of dinosaurs is described using the phylogenetic system. The reference point is the most recent common ancestor (MRCA) of non-avian dinosaurs. Dinosaurs as an entity would be described as the MRCA of triceratops and all its descendants and modern birds. This definition assumes that birds, after they diverged, were still dinosaurs.

Chapter Fourteen

The Cretaceous Period:

Radiation of Flowering Plants; the Cretaceous-Paleogene Extinction

144 mya–65 mya

Elohim pondered on His creations so far. He would soon start to work on producing a human being to populate the earth. Before He got busy with creating a human, He had one more thing He wished to establish. In all His universes there had been flowers with glorious colors, but for some reason, angiosperms, the plant family that produces fragrant and colorful flowers, had not yet diverged. The gymnosperms were present but had no visible displays of flowers. He decided He would indulge in one more creative endeavor. It was still early in the Cenozoic and the asteroid was expected toward the end of the period. He decided to go ahead and create flowers. This was a process that He would have to do in stages and would need some help. This time He did not even think of using His powers with matter. He just thought about it and had an idea. He would get some of the animals to help Him and decided to use monkeys. He excluded the remaining landmasses because they were sparsely populated and had a drier climate. There were multiple groups of them living in forests in Asia and the small islands around landmasses. He was getting quite involved with them and had taught them how to communicate with Him by making noises and pointing at objects to get their point across. He slowly eased into a telepathic mode, increasing it gradually and using signals to create a form of communication that would be guided by Him.

Surprisingly, they quickly got the hang of His means of communication. He imagined that it was not too different from how they communicated. Telepathy, using pictorial messaging, gave them a little pause, but they trusted Him and focused on establishing a reasonable way of communicating using the pictures and signs. He created a basket for seeds and simple implements for digging and scraping the soil. They followed Him and He showed them what He was trying to do by using instruments to clear an area. He also created a few gardening tools and created some cuttings and some seeds. They knew what the seeds were for and reached for them. While they were munching, He dug a hole and planted one of the cuttings. They seemed very familiar with the digging, so He decided this method would work. After a while He began to create the seeds and find a way to disseminate them over the land. Once the plants were established in multiple areas, even the largest of extinction events would not eliminate them. After things settled down, their seeds would germinate, and they would repopulate the earth. It sounded like an impossible feat, but He would devise a solution.

His solution was to engage the help of the birds and the monkeys. He would only treat the forested areas and the landmasses that had living flora and fauna on them. Africa and Eurasia would be important since they would soon be populated with humans.

The climate of the early Cretaceous was warm and mild. Sea levels were high partly due to the lack of glaciers. The giant continent of Pangea had begun to split apart in the Triassic when it divided into two continents, Laurasia and Gondwana. In the Jurassic, more fragmentation occurred and by the early and mid-Cretaceous, this process was completed, creating five continents. Their shapes were similar to the continents of today, but they were not geographically aligned as they are today. As they separated, they collided with each other, initiating earthquakes and volcanic activity, forming new mountain ranges. The mountains effectively removed mass from the oceans, lowering sea levels. Meanwhile molten lava leaked into the oceans, filling deep-sea trenches with spillover of water on to the land. Overall, the result of these opposing actions gave rise to a high sea level but not as high as it could have been. The rise in oceans was significant enough. For example, most of Britain was underwater. In the depths of the oceans, anoxia led to the death of shelly creatures, and their skeletons disintegrated and formed an increasing level of sediment which would eventually be present in our time as the chalk cliffs, as they were called in Darwin's time. Now they are known as the White Cliffs of Dover in Great Britain. The mild climate and elevated sea levels had many positive effects. On the edges of the land and in the new warm seas many marine animal species evolved. The land was covered with forests again. In fact, there were even forests at both poles.

Elohim did little to interfere with the slow pleasant roll of the years that characterized the Cretaceous. He noticed that there were some pre-mammalian monotreme-like species and similarly some placental mammal-like creatures. He did not expect them to survive the collision with an asteroid that had the energy release of a one-hundred-megaton atom bomb. He could have deflected it or stopped it in some way, but He felt it would be a quick and relatively painless end for the dinosaurs. He looked forward in time and imagined the impression their fossilized remains would have on man. But at this moment, the flowers were His biggest concern.

Elohim's presence can be anywhere on the earth instantaneously, so He did not worry about being with the monkeys. While they worked, He transported them to several sites in each landmass that were usable. First, He decided which areas would be prioritized. Then, He needed to make sure that the local weather conditions at each site were not going to kill the plants prematurely. He decided to fabricate ten varieties of flowering plants. Some would derive from seed, while others would derive from cuttings. He included a mix of wildflower seed for the rocky areas to become groundcover that flowered on His initial attempt. He could always repeat it after the effects of the asteroid were gone. Elohim decided to leave Australia's landmass and the Americas out of His first attempt at creating flowering plants. That left Africa and Eurasia as the continents to bring His idea to fruition.

Next, Elohim visited the habitats of the monkeys in several areas where they lived. Africa had several populations, as well as Indonesia. Western Europe had no monkeys. Next, He talked to birds at some sites in Europe about how they felt about planting seed for Him. They were a little hesitant at first, but when He mentioned seeds, they were all for it. After a couple of different bird collection efforts, He had it planned. The birds would simply fly over specific areas and let the seeds fall from a bag with holes in it. This would be especially suited for groundcover. He would leave a separate pile of bird seed from which they could eat. He contacted the birds that lived closest to forested areas and telepathically communicated with them His idea. They seemed quite excited at the prospect of getting seed. He moved forward with His idea, making the perforated bags filled with different seeds and left them at the first location. Wondering if His theory would work, He remembered that birds carry seeds to their young and wondered if they could just take a mouthful and drop the seeds and scratch them into the ground. He would watch the birds and their seed dissemination idea to see if it worked. By flying in concentric circles, the seeds should get sprinkled evenly on to the soil. With multiple birds involved it would not take long if small areas were used. The first flight was rather disastrous; the bag opened before the bird arrived at its location. The next try, He provided the seeds in an open container. They seemed happy with this process, so thirty or so birds volunteered, while others watched. It did not take long for the birds with large claws to scratch the ground at Elohim's request. He thanked them and left a container of seed for them to eat.

Next, He talked to the first group of monkeys about planting cuttings and taught them to lightly sprinkle soil on to the seeds. This was much easier, as they understood immediately what was required and it would be easy for them to accomplish. He advised them to just throw a thin layer of loose dirt over the area where the seeds had fallen. They were excited, as it sounded like a fun activity. He gave them a bowl of seed and bags of dirt and watched as they performed their assignment. They executed the job successfully. In fact, one group sprinkled the seed, and another group covered the patch with dirt. This was very efficient, and Elohim left them fruit as a reward. Once the several groups had carried out their task and been rewarded, He caused a gentle rain to fall over the planted areas. He repeated short bouts of rain every five days, then He waited.

A week went by and most of the areas had tiny baby plants. Now, it was a matter of watching them grow. The soil of the earth was fertile and as long as it did not get too cold the plants would do well. This was especially relevant in areas of Africa which were warm at the time. Elohim enjoyed planting seed this way. He could have the whole earth blooming with flowers just by using His power, but it was much more interesting using the birds and monkeys. It gave Elohim a feeling for what man would experience when seeding land to produce his food. It was also pleasant laboring jointly with the animals and they enjoyed it too. He must remember to teach man that toiling over the soil to produce food and doing it with other people can be a very uplifting experience.

Elohim thought about the process that would occur when the plants matured and flowered. He would have to make sure that the bees and butterflies would be around to continue the cycle. He would arrange for that to happen in each vicinity He had chosen. Since the soil of the earth was rich in nutrients, He did not think fertilizer would be necessary.

The next step was to optimize the frequency of cross-fertilization versus self-fertilization. Sometimes insects will end up introducing the plant's own pollen into the ovary. When that occurs, it is called self-fertilization. The seeds will produce plants exactly like the parent because the genome is composed of both male and female parts from the same plant. Sexual fertilization was introduced by Elohim at some point in the Cambrian after the introduction of meiosis. Now again, Elohim performed the supernatural and produced a plant that instinctively knew how to prevent fertilization by its own pollen. One of the methods was to keep the pollen within its anther at the time insects were visiting the flower. The flowers also tried to selectively attract insects to themselves and to accomplish this feat they used attractive pheromones or colors in their petals. Another idea to attract bugs was to make available a sweet-tasting nectar so that they would choose their flower to feed on. He invented numerous behaviors to both attract vectors to transfer pollen and to force insects to participate in pollination. One example of pollination is described below.

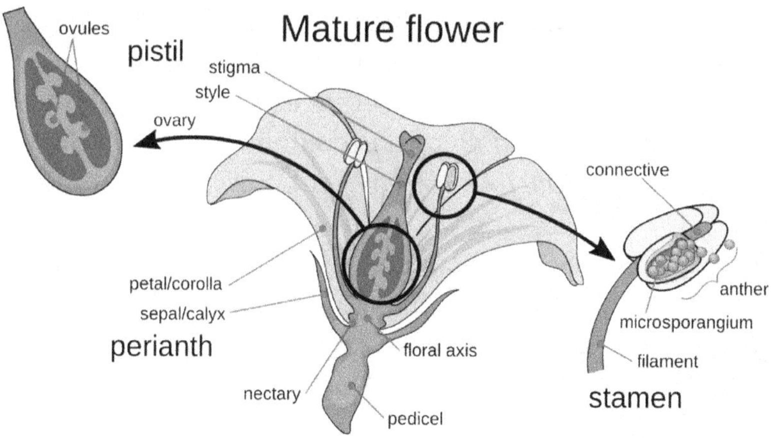

As a sidetrack, one of Elohim's favorite scenarios for achieving pollination and fertilization is a plant called the cuckoopint flower, which has the scientific name of *Arum maculata.* It is a large inflorescence type of flower. It is composed of clusters of small flowers joined together. It has a complex set of actions for attracting insects to achieve cross-pollination. It begins by producing a volatile substance that smells like cow dung. The tiny flies that frequent them love this smell because they lay their eggs in cow dung. They love to collect nectar from the plant stamens. When they arrive at the cup of the flower many of them carry pollen from another arum plant. As they proceed down the flower's chamber, they need to pass through a bristly area of stamens where the pollen brushes onto them. They spend the night in the chamber because it has smooth walls and is impossible to exit. They are happy gorging themselves on nectar, a nutritional substance, while enjoying the smell produced by the female part of the flower. During the night, when the flower's own stamens are no longer active, the stamens generate fresh pollen on to the flies which blends with the sticky substance from the male part of the stem of other flowers they visited, thus accomplishing cross-pollination. The following morning the flies are released by the flower shrinking and retracting its petals. The flies depart, covered in the plant's pollen to accomplish cross-fertilization a second time when they visit a different flower.

There were several methods of preventing self-pollination. One method was to arrange the maturity of pistils at a different stage than the stamens. Elohim continued to introduce variations on this process for different types of flowers.

Elohim introduced hundreds of such scenarios programmed into a variety of plant species. He had a hard time deciding which

He enjoyed more, creating His beloved birds or scheming with the flowers to establish their existence. As it was, the way Elohim created the flora and fauna, the insects enjoyed how He worked the pollination process too, as they ate lots of delicious nectar from flowers. Happy with His new product, Elohim orchestrated interesting triggers that could increase the chance of fertilization. He designed a few insect plant relationships which were very effective, benefiting both partners. Cross-pollinating insects, wind, colors that insects preferred, and scents that attracted them, were all ways to form vector-plant relationships. By the time Elohim was done, He had programmed all the needed instincts in plants and insects. When they were in perfect synchrony, He gave a big sigh of satisfaction. His world was gradually becoming filled with beauty and order.

He had not spent time on the creation of complicated mammalian creatures and kept most of these species entering the Cretaceous, small and still nocturnal. He hoped that man would feel compassion and love for the other creatures in His environment and appreciate their companionship. Even as a predator it was possible for man to limit the killing of other creatures. He should only procure what he needs and not kill randomly. His needs were mainly a protein source as food and animal hides to make clothing to protect against the environment. Lack of necessities could easily be solved in a fair and compassionate manner. Elohim realized that this was only one perspective of the earth. He had created it to provide challenges, beauty, food, water, and all other necessities needed for survival. But there were also unpleasant situations, mostly induced by misbehavior on the part of the inhabitants. He knew His antithesis, Satan, was just waiting for man to evolve so that he could lead him astray and contribute to man's misbehavior.

With no warning, but as Elohim knew, the asteroid plunged offshore at Chicxulub, in Yucatan, Mexico, sixty-five million years ago. Within days the splash was felt worldwide. The only observers were the animals. The expected delayed consequences of such a massive release of energy were noticeable. Elohim had long foreseen this event and knew it would spell the beginning of the end for the dinosaurs.

There was no way to communicate to His animals ahead of time the horror of the event to occur at the end of the Cretaceous. For the past one hundred twenty million years He made life as pleasant as possible for His giant ferocious creatures. Elohim enjoyed watching the great variety of living things that occupied His world. He established a connection with all of them and they sensed Him and were soothed at times of fear of imminent death. As much as He could, without interfering too much with the natural process, He helped them find food and take care of their young. The creatures of the Cretaceous were amazing in the versatility of their physiognomies and their habits. They lived on land, in water, and in the air. Soon He became an accepted and trusted element of their environment. Elohim sensed all their limited thoughts and their feelings. He tried to ensure they did not go hungry and felt as little pain as possible. He helped them cope with hunger and with their sadness when they were not able to save family members. Their sudden change in fortunes was difficult for Him and He tried to keep them occupied with special foods and telling them more often that He loved them. He felt the love they felt for their families, that they shared with Him, and He engulfed them in the all-encompassing compassion of His own. *Would life be like this with man?* He wondered.

The asteroid impacted the earth with a massive release of energy in the form of heat and light equivalent to a one-hundred-million-megaton atom bomb. As Elohim watched, the intense heat that was produced melted rock and formed glass. The massive explosion released chunks of hot rock that scattered widely and started fires all over the world. These caused the atmosphere to be filled with dense clouds of dust and smoke. The energy released impacted all the existing landmasses and caused a severe rise in atmospheric temperatures. The interaction with aerosols in the air and the fires produced vast amounts of soot that accumulated in the upper atmosphere, forming a black canopy that blocked off the rays of the sun. Evidence of this occurrence is still visible as damage was done to rock formations and vegetation.

Scientists think that its impact was felt immediately locally. Within ten minutes, a tsunami probably developed, causing a circular wall of water one mile high that moved outward. In a few hours it expanded eastward and made it all the way to the Indian Ocean. It also proceeded westward going through the seaway that existed at the time between the Atlantic and Pacific Oceans. Once through, it traveled west and joined the eastern wall in the Indian Ocean. This is hypothetical and was predicted by a simulation performed in the present time. Tsunamis are common after earthquakes and volcanic events that affect the seabed, but this is the only one that went around the world. That the tsunami happened is probable, but the extent of it may not have been so widespread. There is no one to tell us what actually happened.

This phase was followed by a steady fall in both atmospheric and land temperatures as clouds of dust and soot in the atmosphere shut out the warming rays of the sun. The lack of sunlight inhibited photosynthesis, and most species of plants were decimated. This affected other creatures in the food chain. At first herbivores died, followed by the predators that fed on them. In addition to reduced photosynthesis, lack of sunshine meant that both infrared rays and photons would be blocked in the atmosphere. In the absence of the warming rays of the sun, temperatures rapidly fell to freezing levels. The lack of light extended to the oceans so that all ocean plants and algae were affected. Cold temperatures resulted rapidly with icing at the poles, causing many cold-blooded species to become extinct. Mammals being warm-blooded were affected less. An ice age began.

Another result of the soot was that it interacted with the ozone in the upper atmosphere, which caused a breakdown in the sky, leading to the production of water that formed clouds and sleet. Shortly thereafter, a steady rain was established planet-wide in susceptible areas. The rain cooled the superheated soot and air, and produced ice crystals which further washed the soot. Unfortunately, the lack of an ozone layer allowed ultraviolet waves from the sun to penetrate the earth which probably affected the remaining life on earth. As the worldwide sleet continued, the soot dissipated and the sunlight began to bathe the earth's surface again. The ozone breakdown stopped and the ozone layer was renewed. As soon as sunlight broke through, all systems reacted to it and the reverse process occurred. Photosynthesis began again, temperatures started to increase, icecaps were reduced, and the ocean and land ecosystems began to be re-established. Considerable new speciation occurred over the next few weeks. Extinctions are one of the main sources of new species formation.

Knowing that the natural processes He had set in place for rapid recovery of environments from an extinction event, Elohim waited patiently for sunlight to return. The rain and ice crystals had been something that He added to speed up the return of life. The time it took for the earth to recover from the disaster is unclear, but recovery from the event appears to have been earlier than scientists today expected. However, experiments conducted today could give inaccurate results because the conditions in the Cretaceous Era when the event occurred are not known for certain.

In the years following the disastrous event, the surviving dinosaurs, which were mostly small ones, increased in number for a while. All large dinosaurs, such as Tyrannosaurus, died in the impact. A thousand years later, a massive release of volcanic gases from Deccan lava vents in India again produced climate change and the non-avian dinosaurs gave up the fight and gradually died away. By about sixty-two million years ago, all non-avian dinosaurs became extinct, as avian dinosaurs continued to evolve into the birds of today.

Chapter Fifteen

The Cenozoic Eon:

Evolution of Primates; Tracking the Flow of the Creation of Animals

66 mya–Present

Elohim introduced hundreds of such scenarios programmed into a variety of plant species. He had a hard time deciding which He had enjoyed more, creating His beloved birds or scheming with the flowers to establish their existence. As it was, the methods and enticements that Elohim created for the flora and fauna that came to the flowers showed that He knew their likes and dislikes and also what made them feel good. They liked His mechanisms for spreading pollen too; it included delicious nectar.

Having traveled with Elohim through three eras of time, the Hadean, the Archean, and the Proterozoic, a pause is needed to reflect on the journey from the beginning of the universe to the present. The Phanerozoic, which extends from 543 million years ago to the present, is the current eon that man knows today. The Cretaceous was the last period in the Mesozoic Era which includes the periods Triassic, Jurassic, and Cretaceous. They were a time of dinosaurs and huge reptiles that ended with a big bang. A large asteroid struck the earth and caused the extinction of many species, including the dinosaurs, which brought creation through 4.6 billion years of the earth's life to the end of the Cretaceous Period. To arrive at the current century, only sixty-six million years of time travel is needed.

The remaining time is divided into segments differently from the previous three. The Cenozoic Era consists of the remaining time, from 66 mya to 2025 CE. The divisions of the Cenozoic were reorganized repeatedly between 1969 to 2009. Currently, there are two consecutive periods, the Paleogene (65.0 mya to 33.7 mya) and the Neogene (33.7 mya to 1.8 mya). The Paleogene includes three epochs: the Paleocene (65 mya to 54.8 mya), the Eocene (55.8 mya to 33.7 mya), and the Oligocene (33.7 mya to 23.7 mya). It is followed by the two epochs of the Neogene: the Miocene (23.7 mya to 5.3 mya) and the Pliocene (5.3 mya to 1.8 mya). The last two epochs were a highly active time in the evolution of mammals, especially primates. This was the time when the placental mammals were due to diverge. It was also the time apes were expected to diverge.

Two consecutive epochs take the place of the old quaternary. They are the Pleistocene (1.8 mya to 0.1 mya) and the Holocene (0.1 mya to 2025). The Cenozoic Era was when most of the mammals evolved into their current state. Elohim has been busy organizing the evolution of mammals. He did not feel that He could leave it entirely to nature, as mammals would play an important role in the life of Homo sapiens. They would become part of his diet, his help with many tasks, and be used as companions that blended with his family.

Elohim decided He would keep track of the groups that formed and moved away and circulated through populations in the various landmasses until they achieved an evolutionary state that He felt was stable. At a point where He could predict the endpoint as satisfactory, He would let nature take over. Getting closer to the current era, versions of animals occurring in the mid- to late Cenozoic could be the same as in modern time. The fauna that exists presently are called extant animals as compared to the ones within the Cenozoic Era. On the other hand, with the Holocene, one might find animals that are still found in the Cenozoic, and they would call the new species extant animals. It is just a difference in perspectives.

As prehistoric creatures evolved, at least a few of them would be in both the Cenozoic Era and the Holocene (these are progressions of time not distance). Scientists in our world are well aware of this situation and take the opportunity to examine live animals or recent fossils and add them to the existing taxonomy. They label fossils according to age and use this information to track the evolution of new species or divergence of existing species with time. New features occur with time. They now know all about the animals and do not have to look for fossils for information about them. Current records allow them to learn when they evolved by examination of their genome. Since writing began in about 3600 BCE, more and more has been documented about creatures that are alive today and are being identified as new species. While on this subject, the first writing that was found is estimated by geologists to be about thirty-six hundred years old. Unless other tablets with writing on them are found, it's assumed that the earliest written history started from that date geologically. This means that after a few hundred thousand years, fewer fossils that are worth investigating will be found. To this point in time Elohim had not created any live animals in either the Miocene or the Pleistocene that could be called extant.

The next chapter outlines the progressive evolution and distribution of various families of creatures over time. Not all creatures will be covered in this chapter—only those documented by scientists as they track successive migrations of some species.

Elohim had been following the family trees of placental mammals since the Cretaceous. As new species diverged, He guided their evolutionary process and kept an eye on them. Present day scientists have charted the sequence of the divergence of different species within a genus. Some of the earliest to be vetted were armadillos, anteaters, and sloths. Next were the hedgehogs, shrews, and moles. Skipping ahead some years, mice, rats, squirrels, and porcupines were created. In the Paleocene Epoch (65 mya to 54.8 mya), a group of primates, which included tarsiers, lemurs, monkeys, apes, and humans, were created. This was an important group since man would be created from one of the lineages Elohim designed. He marked this group as His priority. He had been watching them since eighty-five thousand years ago and knew that as they appeared in the Paleocene Epoch, they would become an emerging family. During the Eocene, whales, dolphins, and porpoises diverged. Another grouping in the Eocene was even-toed ungulates, pigs, hippos, camels, giraffes, cattle, and deer. Later in the Paleocene, a group called odd-toed ungulates, horses, rhinos, and tapirs diverged. There is some controversy as to the exact timeline when some species diverged.

One of the timelines questioned is the divergence of mammals. The writers of this list believe that the divergences of mammals and primates occurred in the Cretaceous. However, science reveals that during the Cretaceous and part of the Cenozoic all mammals and primates were small nocturnal creatures that lived in burrows. Elohim had reduced their size and tempted them into their current living situation. They were aware of the huge dinosaurs that lived among them and were grateful to Elohim for taking care of their safety. If it was not for His actions, they would have, like many other small and medium-sized creatures, become extinct in the Cretaceous. Once the dinosaurs became extinct, Elohim gave the mammals free will to find environments of their choosing and placed no restrictions on them.

Primates diverged from the mammal group around eighty-five million years ago. Elohim decided that He would let them continue to develop and start branching. He expected the first branching to occur in the Paleocene (65 mya) which was soon enough to produce Hominins. They were a very important group to be created, because Elohim planned to use this species as the one that He would convert into early man. It was a pre-human creature that was bipedal and He planned to form them into the first representatives of the Homo species by genetic modification. With Elohim's help, the first human would evolve from this group. He had to make some preparatory changes before He began their creation process. He continued with the evolution of mammals. While doing this, Elohim explored the ancestral origins of monkeys and apes. He needed to see how complicated a process it would be to produce humans using the genetic structure of an ape or a monkey. He looked for potential sources of ape families but discovered that He would have to go as far back as monkeys. If apes had not diverged off the primate line by the Paleocene sixty-five million years ago, He would accelerate their emergence. The creation of man was not too distant.

Prosimians were the first to diverge from the primate family. They are divided into three infraorders: lemurs, lorises, and tarsiers. The lemurs evolved about sixty-six million years ago and soon after they came into existence in Africa, the lemurs gathered together and left the area where they lived and went searching for a better place. Twenty thousand years later, they reached the east coast of Africa. They explored the ocean shore which was swampy with vegetation growing in it. While foraging, they found themselves afloat on mats of vegetation that traveled southeast headed for the island of Madagascar. The ocean current took them to the island in about a month, but food was scarce, and it was very cold. They called on a trait that they possessed, but only occasionally used. They shut down their metabolism to a low baseline and went into a sleep-like condition called torpor. They used the fat they had stored in their bodies for nutrition. It took the lemurs about thirty days to arrive in their new homeland.

Once the lemurs arrived, they found a beautiful tropical environment with many trees and plenty of fruits and nuts. They landed and spread into the forested areas. There were no predators, and they prospered. However, Madagascar was not always a paradise. There were frequent droughts in the dry season and many cyclones. To cope with the environmental conditions, they used their hibernation and torpor abilities and expanded their diet, so they fared better than many other species that lived on the island.

Soon after their arrival, they underwent an adaptive radiation. Lemurs have an interesting morphology. Most have large orange eyes with no consistent body shape or size and have fur in different colors. Examples include the red ruffed lemur: some had black and white faces and looked like racoons, while others had a long fluffy tail with black and white rings on it. They diversified twice more after arrival. The first radiation occurred over a ten-million-year window in the late Eocene-Oligocene. Another occurred during the ten-million-year window in the late Miocene (20 mya). Besides lemurs, the island also had colugos, aye-ayes, and lorises. It is not known how and from where they arrived on the island.

The arrival of modern man around 1500 to 2000 CE had a huge impact on the lemur population. This was mainly related to habitat destruction and hunting by humans. Since then, the overall population has markedly reduced to seventeen species and eight genera have become extinct. The other prosimians left the area to live in different places. Examples of lorises include bush babies and galagoes, which are nocturnal dwellers. They originated in Africa, where many still live, while others are found in Southeast Asia. Tarsiers are South Asian nocturnal dwellers and seem to have ancestral roots in both the haplorrhine and strepsirrhine families. Special types of DNA testing techniques can be used to determine the roots. Due to an unrelated issue, however, the consensus is that they are haplorrhines. All the haplorrhines have a deficiency of a gene that enables the synthesis of vitamin C when present. This means that they are dependent on an external source of vitamin C. This is enough of a phenotype to make them monophyletic haplorrhines. Strepsirrhines have a gene that enables them to manufacture their own vitamin C.

That leaves the anthropoids, which are descendants of one of three precursor families. Parapithecids produced the new-world monkeys which evolved slightly before the old-world monkeys. They wandered off to find a forested area and, after a few thousand years, they found themselves on the west coast of Africa. Interestingly, they suffered the same fate as the lemurs. Only this time they were headed for the coast of South America in the region of the Panama land bridge which was then a feature of the Americas. The monkeys did not have the metabolism-slowing traits that the lemurs had, but their trip was shorter, landing them in Central America which was close to West Africa at that time. The vegetation on the land provided the monkeys with fruits and insects. They found a forested area, and many stayed, but others sought refuge over the land bridge that existed at that time and made their way into the southern part of Central America, which is now known as Panama. The monkeys continued to move south and found lush forests and settled in the Amazon Rainforest. Once settled, they rapidly spread into forest habitats throughout South America. This occurred thirty to forty million years ago. Crossing to North America did not occur. Later, when the warm climate of the Miocene switched to cold temperatures in the Pliocene, they pulled back and occupied the equatorial zone between the two continents.

The old-world monkeys that stayed behind in Africa evolved from a different ancestor called *Aegyptopithecus zeuxis*, the same species from which apes would evolve. They, too, had already evolved before they diverged. Once they diverged in Africa, Elohim saw that they were already a cohesive group and decided not to influence their ongoing evolutionary process. They went on to produce cercopithecines, colobines, baboons, colobus monkeys, and many others. They occupied forested areas, and their fortunes were like those of all tropical species, dependent on the ability to track back and forth between warm and cold areas. If a cold spell caught them by surprise, and they did not move in time, their numbers would be reduced. They did diverge as the earth went into a cold spell which lasted all of the Pleistocene, but since they were in Africa they had just moved to the warmest area. Overall, the genus survived into extant times and are found in Sub-Saharan Africa, Indonesia, and India.

Elohim noted that the lesser apes had not yet appeared. They included the gibbons and siamangs, but they were not due to diverge until the Miocene. Their immediate ancestor, Propliopithecus, already existed so the timing seemed okay. Since Elohim was not using the lesser apes to produce other species, He would wait and see what happened. The lesser apes, gibbons, and siamangs have their own suborder, Hylobates. They originated in Southeast Asia and first emerged in the early Miocene about sixteen million years ago, eventually decreasing in size. The early Miocene was pleasant with warm, moist equatorial forests in both Southeast Asia and Africa. They resorted to a two-to-four ape social grouping which seemed to work well for providing protection and sharing resources when food was scarce, as the Hylobates are tree-dwellers and foragers. There was less sexual pressure on the males too, as they were much smaller than the great apes, which had not yet emerged, so they had less competition. Their choice of very small social groups was unusual for primates, but seemed to be less stressful for both males and females. They had a different mechanism of moving in the trees due to their excellent suspensory structures in their arms, shoulders, and hand joints. Moving by brachiation used less energy than climbing trees or walking on branches, and especially when fruit was hard to find. This gave them multiple advantages. Elohim loved to watch them brachiate, floating from branch with graceful leaps. Being small allowed them to have the full range of the trees and gather fruit on peripheral branches. At that time, they had not left their original home and were still flourishing there.

The late Miocene saw a marked decrease in temperatures and many large apes that left Africa could not tolerate it and either died or moved back to Africa. Apes are broken up into two families, the Hominidae, or the great apes, and the family Hylobatidae, which contain the lesser apes, gibbons, and siamangs. In addition, there were apes that originated on other continents. The great apes originated from the superfamily of African Great Apes and the sub-family called the Ponginae (orangutans). The Ponginae formed their own subfamily which consists of two species originating in two different countries. The superfamily was formed by the African ape tribes which consists of gorillas, chimpanzees, bonobos, and humans. They were supposed to emerge in the Miocene but did not appear.

The orangutans originally occupied a larger territory and were located in Sumatra and Borneo but had also spread to several South Asian countries. They had not yet emerged in the Miocene but their immediate ancestor, Sivapithecus, existed in Sumatra and Elohim felt they would diverge on time. They were not due to diverge for a few thousand years.

It is worth noting here that orangutans are currently not doing well, as the palm oil industry is decimating them. The lush forests that sustain the orangutans are being cut down to plant palm trees. Often the areas have peat bogs, which contain poisonous carbon compounds and cause fires, which are common. The fires also destroy the area for growing forests and kill orangutans that are caught in them. Adult animals are frequently killed, leaving large numbers of orphaned baby orangutans. It is a sad situation, and Elohim is not yet sure if He is going to intervene to save the species.

A taxonomical listing of all the animals that were in the process of diverging is presently complete. To Elohim, the listing was an assessment to ensure that every species was on track to emerge into the twenty-first century of the current era. Other than the great apes from Africa, who had not yet diverged, all other primate groups had diverged and are established in their expected areas.

Chapter Sixteen

The Missing African Apes

The divergence of species continued. Elohim was content that most of the expected species had evolved in a timely manner. There were a few that were pending but their expected divergence was not until the late Pleistocene. He was especially worried about the African apes, of whom there was no sign at all. There were also no signs that they had become extinct. Since He was so close to extant times, He considered proceeding and producing them by using the current ancestors of great apes. He had completed the initial work at the beginning of the evolution of living things. That is how the evolutionary process jumpstarted. He imagined the kind of creatures He wanted to make and designed a few basic body plans that fit them. When they evolved, they would produce more variations and make for interesting creatures. Periodically, there would be adaptive radiations of a new family of animals. He made sure that this did not affect His body plan limit of thirty-six. That is the sequence He used in carefully creating mammals. Since most of the divergence that occurred to date was in the Miocene, He would keep an eye on that time interval.

Elohim thought about where the ape divergence was predicted to occur and saw that most of the primates had already appeared between the Paleocene and the end of the Oligocene. The apes were due to branch out in the Miocene. He found that many primitive anthropoids living in the Fayum area in Egypt could be the used as the ancestors for creating African apes if He chose to produce a few. That was a relief, but He felt the divergence may just have been delayed. Another possibility was that they diverged but promptly left for Europe and became trapped in the ice age in the Pleistocene. There were no recent forms of them, but He knew from glancing at the future that they existed in Africa in different environments. He was now approaching a window of time in which anything He did could change how things were twenty-three million years from the Miocene.

Next, He looked at monkeys. They evolved in time and were well on the way to establishing themselves. He did not have to change anything about them, so He left them alone, but kept an eye on them. He did not think they were likely to produce any unusual creatures, because they already had several species and a great variety of sub-species that were likely candidates for adaptive radiation after they emerged. Some of the apes in Europe and Asia also had adaptive radiations. At the beginning of the Miocene, there were a total of twenty ape species, diverging on different continents. The African Great Apes were not among them. It is important to know the dates of divergences listed in the schedule that was just detailed as information is only one account from many lists, which vary slightly one from another. The lack of divergence of the African apes varied with different theories of what had happened. Obviously, there is no way of directly knowing what happened in real time when looking at the past. The expected dates of divergence have been decided by two different sources of information.

With the advent of molecular clock methods for determining dates of divergence, there have been many assessments of what occurred. The dates are available for many species of animals. Before the molecular clock method was devised, divergences were dated by evidence from fossils. These databases are available to scientists to use for dating purposes, as well as for specific studies. Using fossils looks at the occurrence of a species emerging and having simultaneous adaptive radiation found in the period or a little later in time. Unfortunately, the African apes had always had fewer fossils, and they decreased in the Pleistocene. African apes exist today, though they are an at-risk species.

For the group Dryopithecus, and other local species mentioned before apes, Elohim decided that He could not risk a delay in their evolving since it would affect His timing for creating humans. It took thirty million years for monkeys to evolve from primates to two groups of monkeys with occasional help from Him. This was the interval between the early Paleocene, when primates emerged, to the late Oligocene, when catarrhines evolved and flourished. The difference was that they were a large group with numerous species. If anything had gone wrong with them, it would be hard to catch up. Fortunately, it did not. The apes would be needed in the early Miocene. Five groups had evolved into new apes a million years away from the forward edge of the timeline. He decided that He needed to anticipate the evolution of the great apes. The lesser apes could continue the process they had started, while the orangutans would branch off from their current lineage in about twelve million years.

As a base for creating the first chimpanzees and gorillas, He decided to use an ape ancestor called Dryopithecus. The original ancestor was Propliopithecus, but they were unable to be found and presumed to be extinct. They would not be the same species as the great apes. In fact, many of them originated in Asia and had diverged about ten million years ago. Elohim visited the Fayum region in Egypt and found no African apes, although there were several other species still in existence. He decided to work with the recently evolved Samburupithecus, Nakalipithecus, and Dryopithecus. They had all recently emerged around ten to eleven million years ago and would hopefully be willing. He expected His interaction with the apes would be a little more difficult since this generation were not passive, but curious and interested, and He foresaw lots of questions. He also decided that He would not create any bonobos at this time. Later in the Miocene, when all the gorillas and chimpanzees were successfully produced, He would use one of the newly formed chimpanzees to create a set of cloned bonobos, using two pairs of the new chimpanzees. With time, there would be a good number of them, as they had twenty-five million years to multiply.

He spoke with *Dryopithecus* and asked him for an estimate of the number of his fellow ape ancestors that were of a child-producing age. He was worried that most of the ancestors present were of an older age which would be a problem for gene manipulation. He was not certain but felt He would need at least five couples with a repeat in five years. Elohim was glad

He thought of this possibility, as He would need to use suitable couples if they were willing. He asked *Dryopithecus* to arrange a meeting with whichever couples were interested. Elohim reminded Himself that at least a few new young additions were needed to make sure His species did not become extinct.

The next day, Elohim emerged in the Oligocene in person and met with the Dryopithecus. There were only eight couples, but they were young and of childbearing years. They were all prepared to undergo the process of monitored pregnancies to produce a baby chimpanzee or gorilla. He talked with *Dryopithecus*, sharing His worry that there were too few of them and asked them about the frequency of pregnancies in their immediate group as well as other groups. He agreed that there were fewer babies being born. He also wanted information of the proportion of childbearing females in the entire area but thought He needed to talk to the older generation which would have a better pulse on His question. He told them that He was going to hold off on the gorilla and chimp production until He had more information on how the various clans had grown. He asked the elders in charge if there was any group whose members met at regular intervals to discuss plans for their kind. Most did not know, but one female elder told Him that her father was a member of such a group. Elohim asked her to arrange a meeting with the group. He would have to find a way to enable a two-way conversation and explain His concern telepathically using pictures to show the progressive decrease in the group's membership. Hopefully they were able to return appropriate thought pictures to reply to Him. He told the young couples by signs and thought pictures that He still wanted as many pregnancies as possible to be created. He would ensure the babies' progress in utero was followed and that safe pregnancies occurred. He expected that they would have to make groups for each of the next three to five years. Hopefully the number of females available would be enough that they could become pregnant every two years. He also discussed other measures with *Dryopithecus* for increasing their numbers. Depending on how reproduction progressed, He might be able to create the chimp and gorilla in four or five years. He did not mention the need for the bipedal apes' creation in the early Pliocene because it was about eight million years away and none of them would be around then.

He would see them at several intervals to make sure everything was progressing well. There would be no genetic changes made, as He wanted healthy babies to be born, without modified DNA. The first few babies born were going to be very precious to Elohim as He created His vision for chimps and gorillas. By the time they were in their teens there would be even fewer childbearing members of their species. It would take at least five or six generations to get ahead of the low numbers they had become. Elohim hoped He might be able to determine the reason for this decrease by talking to some older apes. Had they moved the main tribe to a distant place? Or had the low numbers been this way for a while? Elohim asked them to bring their parents and grandparents, and the leaders of their groups to meet with Him.

He asked *Dryopithecus* to be His connection and gave him a gadget with a button on it that he could press to have Elohim contact him. He also asked him to bring together a middle-aged group of two or three of their family members to talk with Him. He would be available the entire next day, then He had to leave. Elohim was worried about the situation. He thought the younger generation caught the seriousness of the situation but did not really know what to expect from the meeting. He hoped He would be able to ascertain the situation when He talked to some adults.

For instance, He would be changing the appearance and biological habits of the Dryopithecus ape babies to those of African gorillas. The babies would be born with some characteristics of gorilla babies and as they grew they would increasingly resemble gorillas as they are known today. The family needed to consider design factors and have both the adult and child Dryopithecus aware of what changes would occur as the baby grew. The babies would have no health problems and would act like any young Dryopithecus—they would eat when hungry and cry when they were uncomfortable. They would have no expectation of what other family members looked like, and they would love whoever or whatever loved them.

Elohim examined the complexity of creating the apes and how long it would take to produce a reasonable number of adult chimpanzees and gorillas. He would worry about the bonobos later. Starting with four chimpanzees and four gorillas, the process would be completed in eight to nine months. He had explained the process to the parents, and they were willing to participate in the creation of a new species. He had one worry—He needed to make sure that enough young apes were left behind to repopulate the ape population.

After looking at all the factors, He decided it was not wise to proceed with producing new apes and that He should postpone it for five years. He could not afford to use any apes to produce the African apes until the local group had increased their numbers. He told the Dryopithecus family that He had decided to cancel the production of African apes and revisit the situation in five years. If He did a limited number of African Great Apes at that time, they would evolve into the creatures they were meant to be. The people who lived in the twenty-fifth century CE would have to prevent them from going extinct. If the current ape populations decreased further, then He would postpone the chimp and gorilla creation indefinitely and focus on the local population. Also, in the Pliocene, He would work on the apes that would need to be transformed into hominins. He would wait until the late Miocene to speak with the elders of that time. They would be an entirely different population by the Pliocene.

Elohim thought about the hominins (bipedal apes) that He planned to introduce. It was a required endeavor. Unless more interaction between apes and early Homo species occurred, the emergence of Homo would be a problem. *The only hope was the planned creation of bipedal apes as intermediaries to humans.* It would be an advance in species until actual humans began to appear. So, the apes had to increase their population and continue to procreate so that their numbers increased by the Pliocene. If Elohim could find ten young participating couples, He decided that would be enough to create the first batch. He would carefully distribute human characteristics spread out between the eight or nine ape pairs. There would be no phylogenetic characteristics. This would help Him distinguish each subtype of bipedal apes in His follow-up of them. He decided to give them a lifespan of forty years each. That way, when He finally created the African apes, they would be mostly extinct. If all went well and the population naturally increased, this would be all He would do for now.

The emergence of apes was not due until the early Pliocene (18 mya). The next day He met with all the Dryopithecus apes that were available. There were eight male and female pairs. He talked to them about the process. He carefully described the problem of their low population numbers. There was no harm and there would be a definite benefit from procreating at an accelerated rate. Unfortunately, they needed to simultaneously address the issue of the missing African Great Apes. To do this, He needed their population to increase, either by more pregnancies or by populations from other areas moving in. They jointly made the decision to go ahead with the eight pairs of apes getting pregnant. They agreed to spread the word about increasing the number of healthy babies being born until the numbers increased. If ten healthy babies were born in nine months, He would encourage more couples to think about getting pregnant. He realized that they had to think about the availability of resources for a rapidly increasing population, but He could support them if the need arose. He informed the elders about this when He spoke to them. Between increased fertility and the potential for increasing the population, their numbers as a species should increase considerably. Meanwhile, both Elohim and *Dryopithecus* would seek information regarding populations of chimpanzees and gorillas that lived in East Africa, Ethiopia, or South Africa.

Once pregnancy occurred, Elohim examined the females and made sure there would be no problems with their pregnancies. This was especially important in first-time pregnant apes. Because of the low population, He wanted to monitor this first group closely. If they increased in numbers by the end of the Miocene (23 mya), and if a trend of a growing Dryopithecus ape population was seen, He could consider the possibility of producing a few great apes. He would wait until the end of the Pliocene to create them since that would leave five million years for them to evolve into extant great apes. Five million years was plenty of time to establish them.

He explained to the apes gathered with Him that the next step would be to increase their ape population. If the pregnancies did well, they would have another group of ape couples try and produce eight more offspring. He contacted the apes at Fayum in Egypt to see if they had seen any sign of the African apes as they journeyed to Europe. This information would help Him make plans. He was informed that there were none, so He placed *Dryopithecus* in charge of contacting the older apes to arrange a meeting with Him.

Elohim told them that He would continue to monitor the pregnancies in person every two months. If any problem occurred between His visits, *Dryopithecus* knew how to reach Him. The goal was to make sure that every pregnancy came to fruition. By the time the babies arrived, He hoped to have looked at all possibilities for the lack or lateness of the divergence of the African Great Apes and lesser apes. The elders left the meeting feeling more confident, and He felt better too. He removed His restraint on the ceiling for fertility of apes as they were barely holding their own.

Elohim looked for information in extant world literature to find out what had transpired with the African apes in the Miocene. There had apparently been an unexplained divergence of apes in the late Miocene. There were suspicions that they may have emerged in Eurasia. Or the apes diverged and soon after left Africa and went to Eurasia searching for food. It was not clear if this included African apes. Fossils for that period are scant, a notable difference from other primates. He thought of the possibility that they may have diverged in Asia and paleontologists seemed to agree. He did find a research paper detailing the appearance of a new large-bodied ape species in Africa.

The study was led by a large group of Japanese investigators headed by Yutaka Kunimatsu. It gave background on the great apes and noted that they did not seem to have diverged. They noted that the fossil finds for the Miocene had been very low for African apes but that they found fossils of several large ape species besides the new species they were describing. It was a new ape called *Nakalipithecus nakayamai*. The fossils were found in two areas: Tugen Hills and the Ngorora Formation. The Ouranopithecus ape fossils date back to nine and a half million years ago, which is close to the projected divergence date for African Great Apes. However, the Ouranopithecus ape was found only in Greece. The investigators' conclusion about the non-appearance of the African Great Apes, chimpanzees, gorillas, and bonobos was that they could have diverged and been killed in Eurasia by cold weather or just not emerged. Elohim was glad to have reinforcement regarding the lack of emergence of the great apes and their occurrence by looking into the future. Trying to imagine the kinds of connections that would be required is mind-bending. It is easier to think of Elohim using His dark matter which has access to everything, than to find the information for Himself. I suspect He did it the most reliable way, no matter how difficult, because that is how He does things. A closing comment in one of the research papers stated that given the presence of the African apes in extant times, it was possible that other species of apes, the Dryopithecus, Nakalipithecus, and Samburupithecus, would activate and diverge at their expected time. Late-emerging African Great Apes would catch up with other species of apes that were emerging and blend in with existing relatives in a future time.

In the current era, African Great Apes are found only in Africa, and they are regarded as an endangered species. They are the same species as the ones that existed in the Cenozoic before they disappeared. The puzzle—how are they in existence now? Several theories exist. During the Cenozoic Era, several precursors of apes existed, and some examples include Nakalipithecus (10 mya), Samburupithecus, Afropithecus (18 mya to 16 mya). Dryopithecus was a recent ancestor of African apes and resembled them in morphology, but none of them are the African Great Apes. Most of these apes emerged eight to nine million years ago as scheduled.

Meanwhile, the two species of gorillas are relatively safe in 2023 CE and will not lack food and care. They live on reserves and the areas of occupation are conserved and occupy the northwest corner of Uganda and spill over into parks in adjoining areas—the Democratic Republic of Congo and Rwanda. These three countries have the extant mountain gorillas, and Gaur, West Africa, has the extant lowland gorillas. Both have been socialized, though extreme precautions are still taken to protect their caregivers and visitors. Each gorilla group is assigned to one human ranger who gradually develops a relationship and interacts with them regularly. Both species are few in number (single digits) and their habitats are monitored by rangers who observe the dynamics of the group. It is unclear to scientists how mountain gorillas exist today when they were extinct in the earth's Holocene. They do not know of Elohim's creation of African apes using apes from other species. He modified the embryos of ten to fifteen apes existing in the earth at the time, by adding and subtracting great ape features to several species of apes that were available from other areas. Their names have been mentioned before. They emerged at a different time from the great apes who did not emerge at all.

Gorillas live in small groups with a dominant male. The male is responsible for defending the female and producing new gorillas. The females also try to mate with a dominant type of male to have progeny with a different genetic result. The females therefore have more protection and more leverage because they hold the key to more children. Gorillas have long cycles between the times that they can conceive so they usually only produce an infant every five to six years. Because of this, when extinctions secondary to cold climate occur, they have a much slower replacement rate than monkeys. Elohim decided that He would wait and see how life progressed for the current apes before starting something with the African Great Apes. He could not afford to risk lowering the numbers of the local apes since they were in the present, and He owed it to them to help them avoid extinction.

The Hylobates include gibbons and siamangs. They were expected to emerge in the mid-Miocene and were supposed to diverge from the Hominidae. There was no sign of them in the early Miocene. Elohim was still in the Paleocene and expected them to emerge when the Miocene advanced. Elohim turned to His next goal which was the emergence of the orangutan. They were due to appear in seven to eight million years. That would most likely happen since it was a long way off. Besides, Sivapithecus, the precursor to gibbons, siamangs, and orangutans, was present so there were available ancestral prototypes if He needed to increase numbers. There seemed to be plenty of them, so Elohim felt that they would continue to evolve and produce the extant orange orangutans that are known today. If He interfered, it could affect the planned process He had already placed for the emergence of primates, and their existence may not happen.

Mammals had speciated beautifully. When discussed in the late Cretaceous, they diversified into numerous phyla and orders. Eutherians, or placental mammals, appeared in the late Cretaceous and continued to proliferate in the Cenozoic. By the Eocene (55 mya), ungulates, pigs, hippos, camels, giraffes, cattle, and deer appeared. Phylogenetic studies conducted on a future earth have shown that most placental orders diverged late in the Cretaceous but extant species appeared later. Fossil studies, however, show that placentals as a group did not appear until the Cenozoic. They also show that mammals were restricted through the Cretaceous and did not speciate until the Paleocene Epoch. The latter is true and fits well with Elohim's protective isolation of mammals in general to avoid becoming decimated by the dinosaurs.

The next epoch was the Miocene (23 mya to 5.5 mya), an active epoch for primates. Elohim had already discovered the problem with the apes and investigated it thoroughly. He decided to wait to produce them. Ancestors of the orangutans (Ponginae), called sivapithecines, spread to western and eastern Asia where they settled and are still established as modern orangutans in the regions. They continued to evolve until they reached the extant form. Elohim thought He might still try to produce the great apes once He saw a progressive rise in population in the other apes. Homo had yet to evolve and was due to diverge in five to six million years. There was plenty of time to get the three species of gorillas, chimpanzees, and bonobos replaced. He made plans to create the species after the production of bipedal apes. The timing would depend on how much the resident ape population increased with special efforts to increase their numbers. The ice age extended into the Pliocene and then into the early Pleistocene and involved all of Northern Europe and Asia.

Elohim watched as once more what was left of the apes made their way back to Africa. They were a bedraggled lot and most looked malnourished. They were glad to return to their homeland, as they had had a rough time in Eurasia. He spoke to them telepathically and welcomed them home. No African Great Apes were present. The prosimians in Asia were doing well, but most lived in the tropical environment of the South Asian islands.

Elohim remembered that He had set limits on the continued increase of new phyla and realized that restrictions could be happening. There were fewer numbers of creatures, especially those with long reproductive cycles. It was a relief that He did not need to expand the body plans. Speciation would continue to provide needed varieties. He had one recourse He could try. He restored the normal physiology for mating so that the few fertile apes that were in existence could multiply.

Chapter Seventeen

Bipedal Apes and Australopithecines

5.3 mya–2.6 mya

As the Miocene Epoch ended and the Pliocene Epoch began, Elohim started to plan how He would gradually transition apes to early hominins. The most recent headcount for the various apes in residence had improved. The first batch of babies had done well in their creation and development. He encouraged pregnancies in younger members of the clan, having them aim for a total number of twenty to thirty for the South African group. He would do a head count in East Africa and Ethiopia and if they still had a significant older population, Elohim would try the same process He had used with the South African apes.

Around 5.3 million years ago, Elohim began making plans to introduce bipedal apes or hominins into the African population to see if they as a group would mate with apes to produce Homo. He decided that He would do this by transforming apes into bipedal creatures with variable Homo-type derived features. As of yet, no Homo-like species had evolved. It was too early in terms of time to begin the process of inducing genetic changes to produce specific anatomic features. The bipedal apes would not be needed until the Pliocene when they were supposed to emerge. He was in the Miocene; it would be twenty-five million years before the Pliocene started.

He only needed to introduce a small number of the species, so He decided that the easiest way to do this would be to travel into the future to the Pliocene and carry out the procedures there. If there were chimps in the Pliocene, His job would be an easy one. Converting apes to hominins was much easier than creating chimps and gorillas from scratch. Traveling in time is not easy but to Elohim, who *is* all of the universe, it is not a problem. It would be a different group of apes that He did not know, but He did not anticipate problems. If there were no African apes but other apes were present, He could recruit them to become hominins. That would be an easy task since hominins would be a newly found species with no affiliation with another genus. He could improvise whatever He wanted and would create several hominin species each with different derived features.

The next day He appeared in the Pliocene. As soon as He arrived, He looked for chimpanzees or gorillas and could not find any. He assumed that they still had not emerged or had appeared and disappeared into areas that they thought might be warmer and more productive for foraging. He decided to use the other apes to produce the hominins. He did not know anyone in the Pliocene. He noted that the other apes were flourishing as there were large collections of them, and appeared to be the same species that were in the Miocene.

His plan was to create bipedal apes in all the inhabited areas of Africa. He would start with South Africa since this was where He spent most of His resting time. They would be programmed so that human traits would be added at intervals of two to three generations. He would check on them periodically for progress. The first batch, though, would be produced by Him as a test of the procedure using volunteer apes.

His plan was to create a population of bipedal beings by introducing genetic instructions for skeletal changes during the development of an ape embryo. He communicated with all the parents, and they were willing. It was very similar to the plan He had discussed with them to create African apes. He discussed the plan with them in several geographical areas that had social groups with which He was familiar. These locations were mostly in Ethiopia, East Africa, including both Tanzania and Kenya, and a few in South Africa. He expected that the skeletal requirements could be induced easily and that once an initial population was achieved, an increase in the prevalence of the traits would occur by sexual reproduction. Both males and females would be carriers of the various genetic traits.

With a small population, this would be rapid, especially since it would give them an advantage in food gathering and hunting small animals. The ability to walk or run at a speedy rate would help them with searching for food, running away from predators, and getting access to peripheral branches of fruit and nut trees.

Assuming only young apes were selected for induction and the frequency of sexual activity at estrus was normal for apes, Elohim calculated it should take a hundred years to produce the number of apes that would be needed. The inability to produce more was affected by the fact that apes could only produce a pregnancy every five years and that they would have to wait until they were adults to accomplish any effects. The local apes stayed in Africa that summer. He would provide food resources when the cold spells occurred. This was a great idea since it solved the problem of reduced apes in various groups in Africa and induced any that diverged to stay put in Africa once cold weather began.

He decided to produce an initial population of ten individuals per target area. In each of them He would introduce two different skeletal features or Homo-type features that could be induced genetically. This would mean ten to twelve traits spread over fifty individual apes using both males and females, a total of two hundred fifty apes every five years. If the traits worked, He could repeat another lot in five years. He should know at that time if it worked.

Two hundred fifty African ape adults should be enough to start the population of bipedal apes. Once this initial population was primed with genetic information, they would multiply in number and the traits would be inheritable. He also added two more Homo-like traits to initiate the process of starting a new species of hominin. Each ape would receive two or three derived traits, but not all the same type. He had a list of relevant features such as foot realignment and the elimination of body hair to help with heat tolerance. He made sure that the apes who were going to be hairless would be alright with the modification. Elohim explained the benefit of this physical characteristic. If they were a couple, He asked the female if she wanted the same. Most would have a gracile appearance. He realized that they would all be receiving overlapping genes but hoped that their basal features varied enough that they would not look identical. Humans had far more physical variables that made up their identity, so it was hard to transition identical-looking apes to humans. He would need to see the babies when they were born, and He told the parents that He would be back at that time. In the meantime, He would keep an eye on things using His dark energy. He made appointments for the intervals when He would have to insert the DNA during the pregnancies. Since most of the pregnant women were around the same stage of early pregnancy, He would visit each site at the right time and perform the insertions.

The apes knew how to contact Him if they had any problems. He would conduct surveillance from the Miocene or whichever period He was visiting. If after two visits everything looked like it was on track, He would let them continue with their lives. It would be a while before the babies showed any signs of their genetic modifications. He looked forward to a group of bipedal creatures with Homo features the next time He saw them, which would be in the middle to late Pliocene. They would have become bipedal creatures with a variety of appearances with no robust features. The population undergoing the process was told that changes may occur in their babies without Elohim doing anything to them. He would try to build a resemblance of their parents into each baby and they were advised to choose features that were represented in their family members.

Elohim prepared the necessary body plans that would allow Him to insert the initial genetic features into the development process. He then talked with each family group consisting of the pregnant female and her mate. He explained that improvements had to occur through the babies by changing the way they grew inside the mother. There would be no danger to either mother or infant—He would make sure of it. He reviewed the kind of changes that could occur and the advantages for them as a group. They would not see a difference immediately, but in a few months, variations in the offsprings' gait would be noticed. In the long term, the whole family would benefit because as the traits spread and more apes became bipedal, they would have more food-gathering options.

The first area He tackled for His follow-up visit was to make the insertions at a site called Aramis in the Middle Awash region of the Afar Triangle in Ethiopia. It looked like the relationship between the group and Elohim continued through the ages. He did not know any of them, but they all knew of Him. He visited the region as a follow-up from the Miocene. He looked for volunteers and asked them to come to a suitable meeting place to review the details of what was to take place. The next day He found about twenty couples that were pregnant waiting for Him. He explained to them the plan and what to expect. The community was quite familiar with the plan and the follow-up schedule. He shared with them the good results of the previous group and told them that the process might happen faster this time. He asked them to return the next day and He would be prepared to do the insertions at that point.

The next day He went through the process of inserting the required genetic code into the embryos. He knew their clan would participate and become part of that new family. They would be intimately connected to Him, and He would keep them comfortable during procedures and remain in contact with them after the process. Elohim was confident that if a real problem occurred with an individual ape, He could fix it and revert the process back to their previous normal.

Elohim started the genetic implantation in a pregnant female ape. She would be used as the prototype. At the correct time of her pregnancy, He inserted the genetic instructions to change two or three features in her baby's body plan. This would allow skeletal changes to the embryo and produce a bipedal gait. Some other steps might be needed after the baby was born and a little older. He explained the process as He worked on her and the results He anticipated. This was now a multi-generational relationship. He could tell that they appreciated His careful procedures and explanations. They were also grateful for the precautions He took in response to their low populations. They had a healthy population at this time, and it was interesting how, even with a primitive culture, they were able to pass important events down through generations. In fact, Elohim sensed that they loved Him. If apes could have a concept of a God figure, this would be it. He yearned for this to happen with His relationship with man.

The babies were a mixture of sexes. As before, He would closely monitor their progress and adjust their development accordingly. It would take a hundred years to make each consecutive change and ensure the process worked. With each new batch of babies, He would check to see what was successfully implanted in them and add any additional genetic modules as needed. The full effect would not be visible until the young apes grew to be adults. He felt that four to five generations, or a few hundred years, would be involved before a population that was uniformly bipedal would emerge. However, He didn't need a large population. If twenty of the subjects achieved bipedality in each of the areas, there would be enough of a population to interact with the Homo species when the follow-up occurred twenty years later. He might not need to create any more if Homo species had appeared. If their numbers had fallen for various reasons, He would to do a smaller batch. The problem was the narrow window for eighteen- to twenty-year-olds that could be used in each sample. As the apes grew older, they might not interact freely with other apes because of having regular partners. The chances of them even interacting with Homo species was unknown.

He planned surveillance from a distance and would know before they returned whether any changes had occurred. The babies would be born around the same time since the pregnancies were all at an early stage. The same process would follow for the ape populations for the other areas. When the subjects were the right age, Elohim would correct the divergent big toes and make them straight, which would be a benefit for their walking. He went ahead with recruitment in the various areas and garnered support in Awash, thanked them for participating, and promised to stay in contact with them. Their experience with Elohim so far told them this would occur, and they did not worry.

Still in the Pliocene, He approached the community in the Taung quarry in South Africa. He used a hologram of a father-like figure to interact with them. He had always been close to the South African group and had seen them before, so He educated them on the process and what it would achieve, especially if their environment changed from forest to savanna. They agreed to participate. He advised them that He would contact them once the results of the first batch of babies were ascertained. If they were favorable, He would offer them the same process. His plan was to follow the first set of hominins for one year to see what resulted. If the traits were accepted, He would visit them every ten years. If the percentage of those who achieved an erect gait was more than 25 percent, He would accept that as an appropriate population to provide at least ten or twelve bipedal apes who were adults.

Ten years later, Elohim followed up and found that half the apes had several features of bipedality. Elohim started receiving reports about the results that occurred. Of course, He already knew since He was keeping close track of them through His dark energy information system. He went to the site to see the new breed for Himself.

The creature He saw was small and the size of a chimpanzee. She was three feet eleven inches tall and weighed one hundred ten pounds. She had a small skull, and her hip-femur relationship was perfectly aligned for a fully upright gait. She now belonged to a group which would be called Ardipithecus. She still had divergent big toes that affected her gait. Elohim began the surgical manipulation of the big toes, which required several years to align, so that they pointed directly forward. This would be His last session with her. He had created a gene for changing the ape anatomy to have the big toes align with the other toes and He would start inserting it into the next few embryos. It would take a while to trickle down to all hominins.

The new creature still had long arms and long fingers to enable her to climb trees. He had not enlarged her brain. Elohim decided that her life would be improved by her upright gait once she got accustomed to it. It would be a few thousand years before climatic changes would occur and her descendants would benefit. With every visit He saw more apes that He had not genetically manipulated become increasingly bipedal. He decided to release her back into her clade to contribute to an increased upright gait in her relatives via sexual reproduction.

When Elohim traveled to the other group's regions, they all wanted to participate. With such excellent results He went ahead and treated the other groups. He would check on them periodically after the insertions were completed and then at twenty years of age. By that time, they would be adults, and He could determine if any of their new children had inherited the traits. He used many subjects in different areas to obtain as much exposure as possible to a larger body of apes. This would produce more variety and more progeny to spread the genes. However, He did not want to increase the number of high-functioning apes since His plan for the hominins was simply to bring about the emergence of Homo. By providing progressively increased advances in appearance and function He would either cause the emergence of the first Homo or would induce mating with any that emerged, resulting in a reasonable basal creature that He could shape into Homo. He hoped this would occur before the hominins, whose scientific name in extant times is australopithecines, all died because of the low longevity He had programmed into them. As a group, hominins would become extinct in about three million years. Until then they would live happily for their apportioned amount of time and then return to His substance.

The advantages the new physical advances gave them were multiple. Obtaining food consistently improved because their gait improved, thus giving them the ability to forage a wider area. They could also see above the increasing prairie grass that now characterized their foraging area and detect enemies at a distance. He expected that there would be new useful traits between the different groups when He surveyed them again in another thousand years. The main purpose of their creation, however, was to allow a collection of desirable traits in one small population and make sure they were exposed to other populations of early Homo species so that the exchange of genes could occur. Their longevity would be approximately fifty years.

The test populations settled into new routines and enjoyed the benefits of walking and running instead of climbing in the two areas. The South African hominin was later named Australopithecus (Au.) africanus. The prevalent species in Ethiopia and East Africa was called Au. afarensis. They emerged a million years apart, afarensis being the first. Each species survived for a million years and then became extinct. They overlapped for about four hundred thousand years somewhere in the middle of their respective million years. The two species lived in different regions in Africa. Afarensis species lived in East Africa, Ethiopia, and Kenya and coexisted with other species. They are considered the members of a clade. It includes Au. garhi and Au. anamensis. Au. africanus lived at the tip of South Africa and the clade included Au. sediba, which was a recent fossil found in extant times.

At the end of the Pliocene, Elohim would decide whether to start the next phase of increasing brain size. Having completed His preliminary group of creatures, the hominins, He was ready to start work on creating early humans. He started beneficial traits that could be transmitted to produce Homo. He stopped for a moment to consider that He had produced a new creature that did not previously exist, Ardipithecus kadabba. There would be several groups in different areas and He hoped to get to know them.

The species would soon flourish in the Afar region and establish well-socialized communities. Elohim was pleased to note that most individual australopithecines would probably have normal survival times for their clade, about forty years. He added one or two human traits to each of the australopiths and also gave them a variety of personalities. They all were calm, pleasant creatures who responded to problems by searching for solutions. They lived as bonded pairs and were involved with their neighbors. They had good social rules and took life one day at a time. He was happy to have created such creatures and hoped the traits would carry over into the rest of His creation, especially Homo sapiens.

Chapter Eighteen

Bipedal Apes and the Fossil Trail

4.1 mya–1.7 mya

The following dates of origination for the australopiths are estimations made using a technique called the molecular clock. In fact, they cannot be configured into any form of natural groups such as a clade or a monophyletic group because they are a new species which are more advanced than apes, and they are bipedal. They have no phylogenetic relationship to each other. This does not include Ardipithecus. Three species are not australopiths and seem to have lived two to three million years earlier than the rest. It is not certain that they are hominins. It is probable that they are not australopiths. The non-australopiths are Sahelanthropus tchadensis, Orrorin tunganensis, and Ardipithecus ramidus. They may not be hominins. They appeared six to four million years ago.

The Pleistocene (5.8 mya to 2.1 mya)

Comparative Summary of Lifespans of Bipedal Apes (Hominins)

Ages of Hominins:

Ar. kadabba – 5.8 mya to 5.2 mya.
Not clear how long they persisted.
May not be a hominin.

Ar. ramidus – 4.4 mya.
No information on when they became extinct. Same species dated to be 4.4 to 5.8 mya? Non-hominin.

Au. anamensis – 4.1 mya.
Disappeared at 3.9 mya.
Interval present 0.2 ky (kiloyear).
Region Kenya.

Au. afarensis – 3.7 mya to 3.0 mya.
Interval present 0.7 mya. Tanzania.

Au. bahrelghazali – Probably a sub-species of Au. afarensis.

Au. africanus – 3.0 mya to 2.1 mya.
Interval 0.9 mya.
Taung, Sterkfontein, Makapansgat, South Africa.

Paranthropus aethiopicus – 2.7 mya to 0.3 mya. Interval 2.4 mya.

Au. garhi – Approximately 2.5 mya.
Bouri, Ethiopia.

Paranthropus boisei – 2.3 mya to 1.4 mya.
East Africa and South Africa.
Interval 0.9 mya.

Au. sediba – 1.95 mya to 1.78 mya.
Malapa, South Africa.
Interval 0.17 mya.

P. robustus – 1.8 mya to 1.5 mya.
South Africa in three areas.
Interval 0.3 mya.

The first hominin fossil is limited to a brain case and is believed to be seven million years old. Found in Chad, it is felt to have been a hominin because of a forward-lying foramen magnum. It is called Sahelanthropus tchadensis. It's possible that it was not a hominin. The next fossil, Orrorin tunganensis, has both a cranium and a postcranial skeleton which includes the spine. It was a bipedal and is six million years old.

There are two species of a genus called Ardipithecus. Ardipithecus kadabba appeared in Ethiopia 5.7 million years ago, while Ar. ramidus, the second species of the same genus was found in Kenya 4.4 million years ago. There are about one hundred fossils of Ar. ramidus which appear to have been well established. Despite being a more recent skeleton, it does not have many derived characteristics and may not have been fully bipedal. These four species are called pre-australopithecines. It was a very gracile genus.

The next two species clustered around 3.9 to 3.0 million years ago. Then came a series of short-lived species of the two genera: Paranthropus and Australopithecus. Paranthropus was robust in cranial and postcranial features but had more derivations that appear to be recent. It was a late arrival on the scene. Other than Au. sediba, the other australopithecines also were on the robust side. Most australopithecines were short-lived and either disappeared or became extinct around forty years of age. The exceptions were Au. afarensis and Au. africanus. Each species survived for about eight hundred thousand years. The entire genus became extinct three million years ago. The lifespan of individuals was about forty years and most disappeared without a trace. The average span of the intervals that they were visible is from 0.17 million years ago for Au. sediba to 2.4 million years ago for Paranthropus aethiopicus, who seems to be the initial species in the South African clade. Some disappeared after a very small interval—0.3 million years ago for P. robustus and 0.9 million years ago for P. boisei. This timeline could change if more fossils are found.

Elohim checked in on His apes at the West Turkana site and was impressed with a clade that had several new beneficial traits compared to Ardipithecus. A species labeled Au. afarensis began to occupy territory in East Africa and Ethiopia. It inhabited areas in Hadar, Ethiopia, Laetoli, Tanzania, and Lake Turkana, Kenya. They seemed to be thriving. Derived features included a flexed skull base, inner ear adaptation to enable fast running, thick enamel on molars, efficient grip function in the hands, a lack of honing complex, and they were fully bipedal. A honing complex is a set of teeth on each side consisting of an upper canine and a lower premolar.

The canine is enlarged, and its tip overlaps the premolar. In robust species, the canine is kept sharp by rubbing against the premolar. In this group, the honing feature was absent and most of them did not have enlarged teeth or jaws because they were eating soft foods. In general, lack of this is an indication of less social aggression in a species or the availability of a soft diet. None of the hominins had aggressive characteristics. They also had an improvement that humans do not have today. Their ilia (pelvic bones) had a lateral flare that increased pelvic stability. Primitive traits included a small brain, and Elohim decided not to perform any modifications, as they were a short-term interim species. He did not want to start new genetic lines based on them. He would give the species a little more time to evolve and check to see if any of the locations had a genus that had developed advanced Homo characteristics.

Next, Elohim checked to see how things were proceeding in South Africa. He was surprised to find that a new Australopithecus species, Au. sediba, had appeared to have evolved 1.85 million years ago. It looked promising to provide access to Homo species. It already seemed to have had some exposure to Homo species and had Homo-like derived features that it could share with its companions. Features also included a thorax which was less flared than other Au. species. It had modern vertebrae with the transitional vertebra from thoracic to lumbar being at T11 instead of T12. The twelfth rib was still attached to the T12 vertebra, providing spinal flexibility. It had a small pelvis with mostly primitive features like a small femoral head and long femoral neck. This lack was compensated for by enlarged sacral bodies and more pronounced sacral curves. The hands showed features that would enable toolmaking, very similar to Homo habilis and Au. africanus. There was a similarity to the Malapa hand that was found in apes in Malapa, South Africa. The brain was small but showed a reorganization of the frontal cortex. It had small teeth and jaws.

Their species spent most of their years in completely different environments. Africanus remained in South Africa where he emerged and afarensis remained in East Africa and Ethiopia where he first appeared. Interestingly, there were three robust species of Paranthropus that have been classified as hominins (some controversy still). They are P. robustus, P. boisei, and Au. aethiopicus. They appeared late in the Pliocene and lasted until the mid-Pleistocene. It would be interesting to see what trails they would leave. They were not good candidates to evolve to Homo because of their robust cranial and skeletal traits. However, Elohim felt He could transfer some of the derived features that were prominent to other australopiths. Paranthropus had not diverged any further and it is likely that they became evolutionary blind ends with no further speciation. Interestingly, *none* of the australopiths gave rise to a new species that was genetically related; Elohim made sure of that.

Looking at the morphological features in the bipedal apes, one notices that until the most recent species appeared, there was little overlap in morphology. The more recent species had more derived features in general, but Au. sediba had the most. The derived hand changes seen in Au. aethiopicus were also present in Au. sediba and Au. africanus. Au. Aethiopicus, who was cited as a probable ancestor of Paranthropus boisei, has some overlapping traits but not enough to confirm a connection between him and Au. sediba. However, Au. aethiopicus could be ancestral to sediba. Despite these changes, mating between australopiths and Homo species cannot be confirmed. Au.

Africanus possessed many Homo-like features but was too recent to be considered an immediate ancestor. This depends on how accurate the dates of their origin were. The caves in South Africa have problems with the usual radioactive dating methods by measuring the rate of decay of the radioactive materials used today because of the structure of the rocks, which interfere with the tracking of the degree of decay of nuclear isotopes in the surrounding rock strata. Alternate methods were not reliable and often increased the possible range by years. Recently a new special isotope was introduced so hopefully more information can be obtained. Homo features, including the brain size, were small in australopithecines but larger in the Paranthropus group.

Derived features seen in Au. africanus were a rounded skull and a large brain with both the parietal and frontal lobes enlarged. Both the sagittal and nuchal crest were muted. He had a human-like dental arcade which included large molars and prominent facial muscles used for mastication. Both features suggest an arid environment and a diet of hard plant material. He also had specialized thumbs that were powerful. Au. Africanus, therefore, like Au. sediba could have been a toolmaker. The increased brain size may have been related to a sparse environment for food gathering. Brains tend to enlarge because of the huge energy needed for chewing hard vegetable material. Au. africanus overlapped a very short time with Au. sediba, so it is unlikely to have been its immediate ancestor.

Contemporary paleontologists are puzzled by the lifelines of the various bipedal apes and wonder why they had such poor longevity and just disappeared early in their existence, one after the other. Times of extinction are not available. They could not find any phylogenetic patterns in their examination of the available fossils. Any similarities that were present were more likely to be homologies, the result of natural causes not genetic ones. Two clades were noted but their similarities were not genetic. There was no sign of a genetic connection between any of them. Numerous evolutionary trees have failed to reveal a relational history between them and it is considered an *unsolved mystery.*

The answer, of course, is that Elohim made no effort to create phylogenetic features in His creations, probably intentionally. He only made a new genus of apes with no genetic connections. Perhaps He wished to leave a puzzle for Homo sapiens to solve. He also did not want them to establish large populations of hominins which could compete with Homo. When multiple species of the genus Homo evolved, He planned to control their numbers as well by reducing their ability to procreate. His plan was to produce a species that was tough enough to survive a hard life initially and, as their circumstances got better, be ready to share their resources and live peacefully with fellow humans. He realized that the chosen species would need to have more skill than the others in their physical abilities, so He would give man an ideal physique.

Without genuine genetic ancestry, each australopith is the beginning of its ancestry and the end. Each species appears to be a single entity with no ancestors or descendants. It only has its own identity. Since Elohim limited their longevity, it is unlikely that any of them had an extended history that can be deduced from their fossilized remains. The Au. afarensis and africanus species survived the longest, each about eight hundred thousand years. The rest of the species were all short-lived or disappeared after a brief period of time for no reason. Fossils of all have been found but are not numerous.

Also, there appears to be no connection between the two species Ardipithecus and Australopithecus. Ardipithecus species multiplied and lived in wooded regions as a communal group. Elohim had already had contact with Ardipithecus before and was impressed with their social habits. They lived in small groups that consisted of bonded pairs and had uncomplicated social communities. It is not a surprise that over a hundred fossils of their kind have been found, including a full skeleton which made news headlines at the time. *Ardi* is a full skeleton of the species Ardipithecus, so there is plenty of material to describe her.

When Elohim created the australopithecines from local apes, He did not know what kind of creatures would emerge. He had created specific species of creatures by inserting physical features using His imagination and made each species slightly different from the others. They appeared sequentially, as He introduced them at small intervals. The result was a group with similar overall appearance but miscellaneous new features. There were no genetic transplants used. Elohim constructed their body forms purely by imagination and used His power with matter to mold what He thought suited their personalities. He used this method of creating to induce whatever He imagined sometimes, as in the evolution of mammals.

Most of the other fossils were distributed in Tanzania, Ethiopia, Kenya, and Malawi. The first fossil was a skeleton of a child they found in Taung, South Africa. It was called the Taung child. Since the South African australopiths were the last to appear it was an early find. This is common with fossils and the story of each one can change when a new one is found. The Taung child was an africanus, which was the most recent hominin species to appear. The mystery is that no traceable geographical, climatic, or other change in environment are known to have suddenly started the emergence of these varied species with sparsity of phylogenetic links. The process does not match with anything that occurred before and paleontologists have not found a reason why they suddenly appeared.

Nor were they the species that brought forth the first species of man. It may seem unfair to them, but while they existed, they had a similar relationship with Elohim that He had with the ardipithecines. To Elohim, they represented an inspirational period of sharing a relationship with a creation that loved Him from the beginning. They could not talk or understand speech, but they knew and felt His presence. They would be an alternate to His angels and He would teach them ways to communicate. He remembered His time in the Cretaceous and His interacting with many species of animals. Even the dinosaurs responded. This would be a much higher level of relationship. He did not feel that they would fulfill their social needs since they were not a genus but just a fixed number of unrelated newcomers.

So, He prepared a new planet for them that already had life on it but no sentient people. Then He transferred the entire population to this planet and settled them on it. He even rejuvenated the extinct sediba clan and some of the other groups that had become extinct so as to add variety to the population. With no other hominins in the creatures that existed there, He hoped that they would form clades of some kind and start producing a new genus. He looked forward to watching over them. Before He transferred them, He took away their sterility and talked to the sediba females that knew about His new plan for creating the genus Homo. He had decided to produce Homo the way He produced the African apes. He would create the embryos for the first ones in pregnant sediba females, so He could control the process at the start and not leave it to chance. He planned to transfer them once their role in producing the first Homo species was completed. He retained the entire Au. sediba clan in South Africa.

He thought of the ardipithecines with whom He already had a similar relationship. He saw that their future looked grim as the forested areas converted to dry savannas in the not-too-distant future. Fruit and nuts would disappear and eventually they would only have dry stems. They would become extinct. He made the decision to transfer them to the planet called Haven, a planet very similar to earth, and let them be company for the australopithecines, who had already been transferred there. He moved the existing population of Ardipithecus. Next, He reproduced those that had died and established them with their kin, as those that had already died left fossils behind. Those that were still young and healthy left nothing behind as traces of themselves. The H. erectus population still had work to do so He did not move them, but would do so later. The sediba clan also remained because He needed them for a plan to create Homo sapiens.

Once He moved the ardipithecines to their new home, He created a safe environment for them and left them alone to produce a new sentient species. He felt sad because of things Homo would do in the future, so He would spend time with these innocents and hope they would evolve into different types of species from Homo sapiens.

His next task was to begin the creation of early Homo sapiens. Elohim arranged to meet with the sediba clan in South Africa and decided that waiting for Homo species to occur on their own was futile, so He talked to them about His plan for producing the species Homo. The group was calm, organized, and not prone to emotional reactions, so they appreciated His method of always telling them what He was going to do. As a group, they were grateful that they had not produced a large progeny. They were the last of the bipedal apes and Elohim had talked to them about their role in producing a human species. They were a recent species and would establish an entirely new family history. Once they were settled in a new environment, they could focus on building a family and a clan. Perhaps the ideal Homo species would come from them.

Chapter Nineteen

Elohim Decides to Create Body Plans for Early Homo Species:

H. Habilis, H. Rudolfensis, and Homo Ergaster; He Adds a Female as a Mate for Ergaster

2.4 mya–1.5 mya

Elohim now set about finding a way to produce early Homo species in a normalized way that included the structure of family. The new Homo species would be sentient at birth, so He did not have to use any of His powers. He had a plan. Before He transported all the hominins and the ardipithecines, He had asked a few Au. sediba women for help. They lived in South Africa and He met several of them when talking to the africanus population. He was introduced to two of them who were pregnant and accompanied by several others who were just curious. He sat them down and told them of His plan. They had no language or speech. So, He conversed in telepathic pictures which He produced in their brains. He managed to get them to reply to a questioning tone with a head movement that meant yes or no. Then He drew pictures in their brains regarding the pregnancy and His making of a better creature that would be more able to take care of his family and his clan.

He would do this by changing the body of a baby while it was in the mother's uterus. This would be done by His willing changes in the embryo of each female volunteer to produce a body plan around which the embryo would evolve. Various automatic segments in sequence would take care of the basic plan. Special features would be added randomly to the basic body plans to produce specific morphologic features. The overall effect would be of a robust head and a postcranial body plan that was very similar to modern humans. There would be three males and one female. He talked to a group of women with their mothers and asked them if they would help Him produce a new kind of their species that would make things better on earth. None would be related to any of the others so each would have a unique identity. They would exist for a small period of time and set the stage for the final creation of man. He had, of course, talked to previous generations in the same way and they were familiar with it. He assured them that no harm would come to the babies. They got together and communicated in their own way and unanimously agreed. He told them He would like a total of four sediba females, three with male embryos and one with a female one. They had to be from unrelated mothers.

The ape families were becoming quite familiar with His requests despite not being the same generation He dealt with previously. Stories of Him were passed onto generations of apes and they respected and loved Elohim. They trusted Him to know what was good for them. The only way to accomplish a new body plan was to create it. He decided to use the young embryos in the Au. sediba females to form a completely new identity—a new species. The anatomy would be that of an upright loose-limbed creature that could walk or run for long periods of time. They would have a larger brain than any species to date and be socially friendly to each other and to other creatures and animals. They would not be His final Homo species but they would be well on the way to it. He asked the two women to talk to their friends who were early in their pregnancy to learn if they wanted to participate. If He had a group of them, He could tell them what sex each baby was and then they could decide among themselves on who would volunteer. They contacted two newly pregnant Au. sediba pairs and asked them if they would be willing to be parents of a totally new and better family of creatures that walked on two legs and were able to talk to each other. They were intrigued by this idea and after some discussion together they were excited to participate. They asked Him what He meant by *talking with each other,* and He explained that by changing certain structures in the neck and within the head He could produce creatures very similar to them that could make sounds that could be interpreted by the others to indicate a specific entity, thought, or event. With this ability a group could share their thoughts, which could be very useful. They liked this idea. They trusted Elohim implicitly, as He was a legend among the apes and was known for His truthfulness and the great care He took to ensure no harm occurred to their kind. He told them to find other newly pregnant couples that could meet with Him as a group. It was important that they were less than two months pregnant, as the changes needed in the babies had to be timed perfectly. He needed at least six pregnant women at roughly the same early stage of pregnancy. He suggested that they ask their parents to accompany them to the next meeting.

Within a couple of months, the women became pregnant, and went to Him as a group to be evaluated. Their mothers were with them. He made sure that none of them were related. He diagnosed the sex of each baby just by looking at their bellies. There were five males and one female. He left them alone for a while to discuss with their parents and each other as to who to choose, as the parents had a say in the decision.

They eventually decided on the four participants. Elohim talked to them trying to explain what He planned to do. This was not easy with the lack of a mutual language. He relied heavily on pictures that He evoked in their brains and He told them of Him making changes by just thinking the variations in each embryo. The variations would start immediately.

They would have abilities that would make them stronger, more agile, and intelligent so that they could take better care of their families. He explained that He would not be inserting anything into the uterus, as He had done with the hominins, but the changes would occur just by Him thinking of them. They would not notice any visible changes for a while but the ability to grow into taller, stronger, and more agile individuals would be sent to the embryo. They would also be smarter. He decided it was time to increase brain size. The Au. sediba pairs knew of His work with the other apes and trusted Him. They told Him to go ahead with the process.

Elohim is within each living thing and knows what is going on in their bodies and minds. He told them that all the children, three males and one female, would be needed to fill the earth with creatures like them, called humans. They would look slightly different but otherwise would be desiring the same things in this life as they did. He asked them if they could accept them as their actual children since they had already been formed in the womb by their loving each other. They stated that they had not thought of them any other way. Also, they had other children so it would not be as hard to release these newly created children when they became adults. They confirmed with Him that the children would live with them as their own once they were born, as they knew how important His work was. He told them He would not have it any other way.

Elohim tried to make communication easier for them. Many things were difficult to explain without language and these were complicated subjects. He went one step further with them and placed thoughts directly in their brain that gave them understanding of what was occurring. He decided that once H. ergaster was ready to do His work, He would give the whole family the anatomy that allowed speech and understanding. Elohim would watch over them and make sure there were no problems. He looked at the three fathers in the chosen families and decided which of them had the closest physical stature for what He would like to see in ergaster. He left the other two to have minor changes toward being human. Otherwise, they would keep their already present genetic makeup.

He thought about the qualities He would induce genetically in the new male infant who was to become H. ergaster/erectus. He would make him an efficient walker and a likeable person. H. ergaster (scientific name) would be tall, with legs longer than arms; his face would be a little rugged, and his personality would be that of an earnest young man who wanted to travel the world and see what Elohim built for him and his species. Elohim would give the child the ability to communicate telepathically and encourage him to respond. He would tell him about the earth and its animals. Elohim knew that the child would one day be using the animals as food, so He would teach him to kill only enough for him and his family and not be wasteful of what they were given. He would share the wonders the earth had for him to explore once he established a clan of relatives and friends to accompany him. Until the human species multiplied, which it would do, they would find other clans of their species in small groups, especially in East Africa.

He had no plans to send them out of Africa but gave them free range of Ethiopia, Western Europe, and all of South Africa, East Africa, and the rift valley.

In nine months, three male and one female Au. sediba babies were born. The boys were named H. habilis and H. rudolfensis and H. ergaster by scientists of today. The female was named by her parents as Eva. The other children also had their parents' names. They lived with their parents until they were in their teens. Most of the traits succeeded—their features were gracile and they appeared intelligent. The hominin He had assigned to be H. erectus was taller and larger than the other males. The males were a little more advanced in terms of derived traits than the australopithecines and ardipithecines, especially in brain size.

There was a chance that H. habilis and H. rudolfensis would find a mate and create more advanced humans—that would be a bonus. Elohim was certain that once the first members of the species formed, it would start a chain of new humans. In fact, H. habilis began to produce simple tools as soon as he became a young man. H. rudolfensis was a reclusive character who we do not know much about. Elohim began to form the most recent male product called H. ergaster by scientists. H. ergaster would be a cheerful, friendly individual and most hominins and animals would like him. His role as a diplomat would introduce humans to less intelligent species as bright social creatures and would not intimidate them. It was essential that early man started existence as a friendly society from the beginning. Elohim knew of the chaos that could occur if the new species began to refuse to share resources with others. Most important was the need for the male product to be an efficient and fast walker and runner. With his companions he would visit every area of the three inhabited continents—Eastern and Western Europe, and Asia. He would cover what he could in his lifetime and his progeny would continue the job. Neither would return to Africa in their lifetime. Elohim hoped by this means to spread derived inheritable characteristics within the populations, slowly inducing improvements in overall appearance and the abilities of the species. Elohim's intent was for the female Au. sediba baby to be the mate for H. ergaster and she would have his level of intelligence. She was not related to him since their parents were not linked genetically. Her parents would raise her as they pleased. He trusted their judgment.

Elohim planned for H. ergaster to gradually evolve into a wise and patient representative who would spread his knowledge and promote an optimal way of life to his fellow humans. In the process he would change both physically and in wisdom. He was committed to do everything that Elohim asked him to do. Elohim loved him the most of all the pre-human creatures He created. He allowed him a long life and his species would last about one and half million years. His plan was to send all of them to the planet of the hominins when they died. They were kindred spirits. Physically H. erectus was created by Elohim, feature by feature exactly as He wanted him to look. Looking at the child erectus in His mind, He imagined that below the neck he would have hairless skin and well-balanced muscles to promote fast, fluid walking. H. erectus would need feet that provided good support, while his hands would need to have sensitive finger pads and a thumb that would enable the invention of new tools. His arms would be shorter than his legs, giving him the ability to boost his stride as he walked efficiently across the savanna that was rapidly replacing the forest. He made sure all the features required for bipedality were in place. He decided to leave the face rugged with prominent eyebrow ridges. An image of strength and power would be important to give him the confidence to meet and greet all kinds of species and introduce the idea of a world where all creatures lived in peace with each other and with their Maker.

Most importantly, Elohim gave him a large brain. He would make his teeth small hoping that he would be able to find softer vegetation and fruit. If his diet at any time changed to hard plant material, the teeth had the capacity to become larger, especially the molars. He needed to supplement plants with the meat of small animals because his muscles and larger brain would need a lot of energy. With his physical attributes and natural bravery, H. erectus would make a good hunter.

He added a face that smiled easily, the most important feature that would be imprinted in his brain. This feature gave H. ergaster the ability to grin easily, radiating genuine warmth and interest in whomever he was talking to. The improved aesthetics and his friendly nature would allow his persona Homo ergaster to travel from South Africa to Olduvai, Tanzania, and from there to Ethiopia. From Ethiopia he would make his way across to the western shore of the Red Sea and follow it northward to make it to the Middle East. Alternatively, they could cross the narrow straits between the Horn of Africa and Yemen of today, and follow the coastline to the Persian Gulf. This would open many possibilities. For most of their journey they would follow the animal trails which were well trodden and would keep them close to food.

For the female He planned only a few characteristics; her anatomy would be like her husband's, making her a good walker. Her personality would consist of bravery and endurance, and she would be a loving person whom her husband could lean on. As the leader of the group, ergaster would need an adviser he could trust. It was important that they worked as a team and set an example for the rest of the group. H. ergaster, H. habilis, and H. rudolfensis would be a genetic composite of the characteristics of several species—Au. sediba, early ape ancestors, and all the genetic programming that He introduced. He had no doubt that when the four new individuals came of age, they would be advanced enough to call them Homo. It would be a rather archaic form of Homo, but it would not take long for them to evolve to the final product that Elohim saw in the future.

In eight months, the three male humans were born as three lively looking Au. sediba babies. H habilis and rudolfensis had early Homo characteristics and were mostly gracile. Rudolfensis was taller than habilis. They seemed to be healthy on a cursory exam done by Elohim. They remained in the care of their delighted parents. H. ergaster was a happy baby and adapted well to his new environment. Elohim felt that it was a good sign. He looked into the future and saw a young teenage H. ergaster striding across the savanna, his head held high and his arms powering through the air to give him additional leverage to move forward. He was robust in face but so were all his co-existing hominin species. He had two advantages in dealing with hot climates. His tall slim body could efficiently radiate out heat with its high skin-to-body-mass ratio and he was hairless, giving him the ability to sweat. Sweating causes loss of heat by using body heat as energy to evaporate moisture and as a result produced cooling of the body as a whole. Hairy bodies trap heat and are better in cold climates. The female was pretty with no robust features. She also appeared to be normal upon examination and slept a lot.

It would be at least another fifteen years before they all grew up and could start the work of expanding the human population. Elohim was content that He had at least initiated the process of evolving the human species. During the next fifteen years He planned to introduce Himself to the four and stay connected with them as they grew. Initially, when they were just babies, He simply reassured and cared for them when they seemed afraid or had discomfort. He also made them aware of His love for them. He learned from experience with animals that it was best to introduce Himself with love and reassurance so that once they started to think for themselves, they would already have a trusting, loving relationship with Him. As they grew older and began to learn right from wrong, He gently showed them the way. He was always there when they called upon Him. Once they could talk, He taught them to call Him "Father." Since they did not have a sophisticated language, He communicated with them in the same manner that He did with their parents, telepathically with pictures.

Up to this point, He had not enabled the physical changes required for efficient speech into any Au. or Homo species. He wanted to wait and see which species were likely to survive. There would be frequent evolutionary dead ends for some. In addition, the creation of man was going to be forged by two lines of Homo simultaneously. H. ergaster in Africa, then Asia, and at some point be replaced by H. erectus, who would be instrumental in working with AMH (Anatomically Modern Humans) from South Africa, East Africa, and Eurasia to blend the different personalities they started with into one well-defined population called Homo sapiens sapiens. H. heidelbergensis in South Africa played a role in merging the clusters of archaic Homo sapiens and AMH coming back from Asia to form H. s. sapiens.

The H. heidelbergensis would initially start in Omo, Ethiopia, and would travel to North Africa (Morocco), then to the Middle East, especially Israel, and then cover large areas in Europe and Asia, accompanying AMH as they wandered into Western Europe, Eastern Europe, and Asia and then back to Africa. Homo heidelbergensis and H. erectus merged with different AMH groups and both were on the move constantly, so much so that people would confuse one for the other. At times they could be in the same area for a period of time and not be aware of the other's presence. The other two males were named H. habilis and H. rudolfensis, which were the names historians gave them in the future. Elohim did not think H. rudolfensis would travel far from his home, while fossils of H. habilis have been found in China and Eurasia. He would probably be independent and not want to be under Elohim's guidance. Elohim decided He would keep an eye on both of H. habilis and H. rudolfensis and be available if they needed Him.

Elohim smiled as He brought up the picture of ergaster striding forward with arms swinging, pushing off with his feet for each step. He adventured his way around Eurasia, setting a good example for man and getting to know Homo sapiens as he evolved. He stopped at intervals in places that had a high concentration of Homo sapiens and talked to them about Elohim. Most likely some of his group would want to set roots in such areas. This would be a good idea once they spent some time as a group and got to know each other well. Traveling across both Asia and Europe gave them some familiarity with geography. Having new friends in different areas helped them when they eventually decided where they wanted to settle. He managed to convince them of the wisdom of this approach, and they were glad for His guidance later.

Elohim was pleased with ergaster and his group. He decided that He would let the species live a long time to balance the more aggressive humans who loomed in their future. Soon they would start populating Eurasia. Talking to ergaster's family, He was satisfied with the living habits and the interpersonal relationships of the Au. sediba clan. He felt ergaster would grow up to be just the kind of emissary He needed to speak to the various intermediary species of Homo that would soon begin to populate Africa and Asia. His family had a roving routine, moving from area to area between South Africa and East Africa, returning to their permanent settlement when the animals and fruit decreased in other areas. They felt their children should learn to be as mobile as they were, which suited Elohim's plans too. The australopiths had a short species lifespan, and then transferred to another planet. As a result, they left only a few fossils of themselves. H. ergaster, followed by H. erectus, continued the work of educating AMH. They did not create any new species that were different from them. The lack of continuity in the few fossils that are available is puzzling and just as we were getting to know them they were gone. There is no evidence that they ever left for Asia or Europe and no hominin fossils have been found outside Africa. So, the lack of genetic continuity would not matter, as they did not know any different.

Chapter Twenty

Homo Ergaster to Homo Erectus Grade

2.1 mya–117 kya

Homo ergaster lived with his parents until he was a teenager. He had comfortably settled into an Au. sediba clan, despite its considerable derived features. He was the son of two sediba parents and for the most part looked like any of his sediba cousins. He had frequent contact with Elohim who taught him that He needed him to explore the earth and establish new families of humans in every place he settled. Elohim would make sure they had sources of food and other needs. He had given him a body that was designed for walking or running in an efficient way, without tiring easily. H. ergaster would find the earth to be an amazing, beautiful place with many animals with whom he shared the earth along with other humans. Animal predators targeted his food but would not hurt him if he did not hurt them first.

Over the years, H. ergaster had many talks with Elohim about his future journey and he felt that the time was near for him to move. He was beginning to get facial hair, which was one of the first signs of growing up into adulthood. From the beginning he felt the urge to travel but did not know why. The sediba clan was well liked, and he had a good life. When he got into his teens, the urge became much stronger, so he talked with Elohim in His mysterious way of communicating telepathically.

The rest of his family could not communicate in this manner. However, two of his friends, whose families were close in physical characteristics, had the same experience with Elohim, but not as often. Most of their kind wandered from their home in South Africa and could mostly be found in East Africa. He missed them and was reminded that he had been selected to explore his world and that he should gather a small group of his friends and family and see what the world was like. But the chances were strong that they would part company when they found a place that they liked.

His physical characteristics were different from the rest of his family. Some resembled him in physique, but their thoughts were limited to everyday events. He prepared for a long journey and packed warm clothing as he heard from others that it could get cold in the northern regions where he was headed. He packed his best tools, especially those that were added to the family group. Up to this point he lived in an area that he knew well. When he reached the destination Elohim planned for him, he would meet species that had completely different lives. His focus was to remember that he was the only one of his kind and needed to produce more of his kind. He needed to ask Elohim and his stepfather how to accomplish the goal. He was of age now and could have a family, yet he was still living with his mother's tribe at their South African homesite near Malapa Cave.

Homo ergaster had robust cranio-facial features compared to modern humans. This may have been diet-based. His brain case indicated a large brain of approximately 1,200 cc. His skull height was lower than modern humans and he had moderately sized supraorbital ridges. He had some postorbital constriction which was common in robust faces and had a human-like projecting nose which provided time for air breathed in to warm up and humidify the nasal lining. He did not have a chin. Compared to the Asian version of his species, H. erectus, he had a larger brain and a longer skull vault. The skull bones were also thinner and therefore lighter. Postcranially (below the skull), he looked like a modern-day human.

The climate of Africa in the Pleistocene was very hot and arid. Homo ergaster was glad to be able to leave the area and look for a cooler place to live. At the time there was volcanic activity in the Rift Valley area and many people from his clan left when the heat became intolerable. He was glad to have the initiative to go and discover other areas.

Before he left, he wanted to talk to *Father Elohim*, as he called Him. He would miss Him and hoped that a connection would be maintained after he left the area. He closed his eyes and called him softly, "Father?" Immediately, the voice responded telepathically, "Yes, My son. Are you ready to leave?" Elohim saw that H. ergaster was anxious about going out on his own, and wanted to talk about a matter that had been worrying him. The boy was the only one of his kind. He had just come of age, and he was well on the way to becoming Homo sapiens. It was important to continue his line.

Elohim planned when He made him to create a second human-like baby except that one was a girl. He had not kept in touch with her parents, and they were unrelated to the parents of H. ergaster so she could be a suitable match for him. H. ergaster would not be returning to Africa and his life would be about exploring the new world and talking to Elohim about the things he had seen. Elohim, of course, automatically tracked everything, as it was important that He knew how the new Homo-like species felt about the increasing competition for resources and to ensure that peaceful relationships existed.

Elohim would also be watching to see if advancement toward the eventual Homo sapiens holotype would occur from H. ergaster's progeny. Elohim asked him, "Have you said goodbye to your parents?" He replied that he had. Elohim asked that he meet Him at his home because He wanted to talk to his parents and him together.

Elohim made Himself present at the parents' house. Fortunately, they were home and greeted Him with a smile and waited. He always had something to tell them when He stopped by unexpectedly. He told them, "Your stepson is ready to leave on his long walk, but there is something amiss and I am not done. You remember the other couple who had their baby treated too? Are they still friends of yours?" The couple nodded happily that they were. "I had hoped that she may be a mate for him. I assume they know each other since your families are friends. How do you feel about this idea, and do you think they will like each other?" The mother's face lit up. "They are good friends, and I think they will both like the idea. Will she be able to travel with him?" She looked at her husband and he smiled and told her he also liked the idea.

Elohim asked them to talk to the two young people together and if they agreed to be mates, their departure would be postponed and a celebration with their friends would be arranged. He did not think their son would be returning because H. ergaster would be traveling to distant places, constantly on the move, staying for short periods in places where they were welcome. The female's parents looked sad but knew that this was their daughter's destiny. Besides, they had also become fond of ergaster and knew he would take care of her. To comfort them, Elohim added and told both sets of parents, "I have looked into the future, and they will both live to an old age." He asked them to have the young man contact Him with a report.

The following day, ergaster, after talking to his parents and his mate-to-be, called Elohim to tell of his plans. Eva was happy with the news that they would be a mating pair committed for life. They got along well and loved each other in a brotherly/ sisterly way. That would change rapidly as they grew up into a loving marital relationship. They had different parents and were not related to each other. Both mothers had had genetic derived traits introduced into their embryos that they had with their husbands. So, they were actually the real children of their parents who would miss them. It was good that the couple already knew each other, having grown up together. It made for the likelihood of a good long-term relationship. Theirs would be the task of enlarging the H. ergaster clan. They were going to have a small celebration and ceremony committing themselves to each other. Then they would leave on their journey. Elohim was invited to be at the ceremony.

The following day, Elohim met the whole extended family. It was not big, but they appeared to be very close to each other. Elohim appeared as a bearded grandfatherly representation just as the parents saw Him throughout the sessions of introducing genes into their daughter's embryo, so they were not startled. They had had a positive experience and loved and trusted Him to look after their daughter and ergaster. Somehow, they knew He could do it.

Homo ergaster prepared to leave his home in South Africa with about thirty of his clan. The clan made plans to make its way first to East Africa and the Olduvai Gorge. There was a collection of sediba at their destination. Elohim gave them directions to East Africa and told them He would be in contact to help them when they arrived at their next location. He told them about locating the North Star because once it was identified they would know how to discover the other directions. For now, He would contact them each morning and point them in the general direction to travel. They would walk via the Olduvai Gorge which was familiar to them, rest there for a while and gather supplies. From there they headed toward Omo in Ethiopia. They had friends and family there and would spend a few months learning more about the next location. Most of the males were joined by a female and most were in their teens or twenties. Two or three older men joined the clan and shared their wisdom and life experiences as they traveled.

They were armed with their Oldowan tools that they made themselves. They made hand axes from stones large enough to close one's fist over. It was then weaponized by taking another hard stone and knocking off flakes around its edge on both surfaces. This provided a bifacially sharp edge. A second step of using a hard bone to strike off a finer flake from the first edges would make it a potent hand-held axe. In addition, he had tools like choppers, flake tools, and cleavers. Some flakes were used for scraping hides clean and others for cutting them. One of the men traveling with the clan was his uncle, who had considerable skills in creating tools and wanted to experiment with a throwable or hand-held staff with a sharp flake at the end of it.

His uncle had recently been experimenting with fire. The first time H. ergaster saw his uncle start a fire he was astonished and fearful. When he was shown how to safely confine the fire between rocks and the process to start the fire by using two pieces of flinty rock next to some dry leaves and twigs, ergaster was amazed. His uncle cautioned him about the risks of fire and how he should dampen it with water, then rake it until there were no embers and the area was cold before he left the site. To be extra safe, water could be poured over it if it could be spared.

Homo ergaster prepared to leave his home in South Africa with about thirty of his clan. They consisted mostly of his teenage friends. His parents had persuaded him to invite two older individuals who were family friends and knew most of his clan. They made sure they had the tools they were used to. It would be a long journey, so they planned to move fast either running or walking until they were tired, stopping only for food and water. They would make stops of two to three days if they found a good area to stock up on fruits and other edible plants and try to trap or kill a small animal for food. The two older sediba males knew how to light a fire and they also had learned about cooking strips of meat on a hot, flat stone laid within the fire. The young, adventurous Homo ergaster loved the experience. So did most of his friends and family. They were excited about new places to see and new kinds of creatures like himself to find. He did not yet feel the great responsibility he carried. He was going to be the vehicle through whom Elohim would bring some order in the formation of a new genus. It would initially have several species but eventually He would choose the one that would remain. The rest would become extinct, one after the other, at optimal times. Homo ergaster would not be one of the candidates. He did not have all the qualities needed, but he would be a loyal and dedicated helper to Elohim throughout his long life. When he completed his work, Elohim would send him to Haven to join his family who had agreed to precede him there with the rest of the sediba clan. They would meet up with the remaining bipedal apes that were already there.

His group set off from South Africa headed toward Olduvai, Tanzania, going in a northeast direction. From there they would head to Omo in Ethiopia. They heard that a waterway not far from Omo could be used if they created floating rafts to cross over to the other side. They would look for bamboo or other wood as they drew closer and use it to make a raft. They had been told by others who left and returned to their homeland that if one followed the coastline in a northerly direction, they would arrive at a land where they could go eastward into a large body of water. It would take a while to arrive, so they fed themselves along the way with fruit, nuts, or hunted for small animals.

Homo ergaster species had the physique to be rapid walkers or runners. As long they were well fed, they could cover many miles. Elohim had already activated the language/speech centers in the brain soon after ergaster agreed to the plan. He had done the same for the group going with him and his family. He talked to ergaster about what to expect so he could reassure his companions when the first signs of having the language/speech centers began to happen. Into ergaster's mate he installed the mitochondrial gene, which would be used as a tracking device for Elohim to use to determine what routes they took and the timing of their movements. He expected all of them to look different the next time He saw them.

This would be largely due to the establishment of speech. Speech, both the production and understanding of it, would help with sharing new ideas with others and building upon them. With the ability to communicate, not only could ideas be shared but tasks would become more efficient. They would perceive the world in a totally new way. Once speech was achieved, language would follow and better communication would lead both to the accumulation of memories. They would have the ability to plan their actions as a group. None of this would happen suddenly. Their abilities and how to use them would slowly evolve until they became a way of life. Then they would be able to spend time relaxing and exchange ideas. Gradually new vocabulary would grow, and a language would be formed.

This was happening in other groups that had received the speech-related changes earlier. As a result, both their group and individual knowledge increased. This would leave time for imagination. Elohim would add to their knowledge if they asked Him questions. The group was a little intimidated by Him, but as time went on, they felt nothing but love for Him and trusted Him implicitly with their lives.

While ergaster grew up in South Africa, other erectus grade species developed in Indonesia, Dmanisi in the Republic of Georgia, and Java in the island of Flores. Homo floresiensis was a pigmy species of H. erectus grade. He established himself on Flores Island and probably traveled to Asia as a full-sized H. ergaster and reached the east Indian islands early in their travels. The species never left the island and became miniatures; although not hormonal dwarfs, they were normal-proportioned humans. They fought miniature versions of elephant precursors and lived longer than most other Homo species, even surviving the explosion of Mount Toba. They are believed to have become extinct about fifty thousand years ago.

The Homo erectus grade species that evolved in Dmanisi in Georgia (Europe) became extinct earlier. Not much is known about it. The transition of Homo ergaster to Homo erectus grade existed in a large area in Eurasia, making fossil identification even more complex.

Chapter Twenty-One

Evolution of Homo Neanderthalensis and Anatomically Modern Humans (AMH)

from an Unknown Ancestor in Spain

700 kya–40 kya

Homo heidelbergensis in South Africa flourished. Elohim had a given them the gift of speech as well. They had the beginnings of a language. He also installed the mitochondrial gene in a female heidelbergensis. That would enable Him to track the group and know where they traveled. The heidelbergensis group in Bodo that was heading toward the Middle East and Europe received the same treatment before they left. The group was established in the Bodo area and decided to migrate. They were a species that diverged from an ancestor that lived eight hundred thousand years ago. The ancestor, which we have not yet identified, is believed to have been advanced with a large brain and intellectual abilities ahead of his time. After they arrived in Bodo, the group split into two. One group was told by Elohim to head south until they reached the Klasies River at the southern tip of South Africa. They are called African heidelbergensis. The other half of the group headed northward from Bodo and arrived in the Middle East where they settled as evidenced by fossils. From there they traveled to Europe, leaving a trail of bones that were discovered, giving scientists an idea of how long ago they lived in the area and what other Homo species with whom they crossed paths.

Elohim noticed a species of Homo sapiens with varying archaic features, but also with many modern features. They are called archaic Homo sapiens. They vary in age from three hundred thousand years ago to one hundred twenty thousand years ago. Found in a cooler climate, African fossils from Europe have viable DNA and molecular clock and genome studies can be used to date their existence up to a million years at discovery. In Africa, because of high temperatures most of the year, DNA rarely survives long after death. So, the oldest DNA that may be viable in Africa is less than ten thousand years old.

One interesting find involving Eurasian heidelbergensis is called the *pit of bones.* It was found in the Atapuerca archeological site in Spain. The site area is a vast archeological enterprise which has lasted for years. The cavern was found in the Atapuerca Mountains when attempts were made to open a new railroad. When they dug into the mountain side, workers found layers of rock interspersed with layers of fossils. The archeological site began in the 1980s and is still active today. Bones were found in a pit in the cave system and have provided a great deal of information. There are several sites focused on different objectives. A huge amount of information on early Homo species has been collected. The skeletal fossils include Neanderthal, heidelbergensis, and even H. erectus bones and skeletons. Large quantities of cultural objects are being studied, shedding light on their way of life. Tools and cultural objects have enabled dating of the fossils. They range in age from three hundred thousand to one million years ago.

One site, called the Gran Dolina, has been studied to a depth of eleven sequential levels. The sequence went from the most recent to the oldest. Each level had different findings. They ranged from early Homo and hunter/gatherer groups at the lowest levels to Bronze Age occupants at the top. The fossils and cultural objects, like tools, provide detailed information on recorded human occupation over a period of 1.2 billion years ago. In 1994, multiple bone fragments of several species of Homo were found. One of the findings resolved a loosely held theory on the ancestral positions of Homo heidelbergensis, Homo neanderthalensis, Homo sapiens, and Homo antecessor. Findings at a second site, Sima de los Huesos, reinforced these findings. The site consisted of a pit at the bottom of a deep shaft containing bones that belonged to Homo neanderthalensis (the bottom layer was dated at over three hundred fifty thousand years old).

The pit contained complete skeletons, one laid one on top of another over a long period of time with the most recent bodies on top. They were all Neanderthal bones. Scientists studied the genome using molecular clock technology to compare it to a modern genome to determine how long it took for parts that were common to both genomes to have changed to the present. This time interval revealed the age of the bones, which ranged from 550 kya to 750 kya (thousand years ago). Although the results are not precise because of the nature of the methodology, it does give an approximate timeline. Prior to the use of this technology the oldest H. heidelbergensis fossils dated from three hundred thousand years ago. In this case, the oldest fossils found dated back 550 kya to 750 kya. From this data, they surmised that the ancestor lived about eight hundred thousand years ago. Scientists feel that Homo sapiens species diverged from this ancestor soon after heidelbergensis did. However, no Homo sapiens fossils have been found at the site.

So, Homo heidelbergensis was probably the offshoot of the ancestor who lived about eight hundred thousand years ago. He was intelligent, creative, and well-liked by all the species he contacted. Physically he was more derived than H. erectus and Elohim chose him to produce two new species in Eurasia: Homo neanderthalensis and the Denisovans. He told the first group of heidelbergensis in Omo, Ethiopia, to make their way to Morocco in North Africa. From there they moved to the Levant. It is possible they spent some time in that location overlapping with anatomically modern humans (AMH). The actual time spent together may have been limited, however, as African AMH migrated during the intolerably hot summer months in South Africa.

Heidelbergensis spent time in the Middle East when Europe was freezing cold. He is also documented to have been in Spain, Arago, Germany, Italy, Hungary, Greece, and China, as evidenced by fossils. He spent some time in Israel as well, evidenced by early fossils that were found there. In Africa, he left fossils in Morocco, Ethiopia, Zambia, Tanzania, and South Africa. In some ways he moved around very much like H. ergaster but had a different route. Movement is detected by fossil finds in various places and by determining the age of each fossil. Elohim, of course, tracked him through the mitochondrial gene that He had placed in one of the female relatives that accompanied him. It is difficult to be sure that discoveries are interpreted correctly.

Some of the apparent movements may have been simply from early fossils being catalogued inaccurately. Early archaic Homo sapiens (EAHS) followed the same paths in Europe as heidelbergensis did. It is possible that they were responsible for the fossils and that heidelbergensis had never been there. It was also technically difficult to date and to identify the species from fossils. At some point a decision was made by paleontologists to reassign all EAHS fossils to heidelbergensis and it is unclear how many of them were really this species. Between H. habilis, H. erectus/ergaster, H. heidelbergensis, AMH, and for a short while, H. antecessor, traveling across all Asia and Western Europe, people often mistook them for each other. Also, after some time, all species developed new physical changes that would make them look different. This was evidenced in the fossils that formed when they died. It appeared that mating between all the different Homo species, even archaic ones, was common, and this led to more variety in their features as they aged. Thus, it became increasingly difficult to know which fossils belonged to which species, as time progressed.

Heidelbergensis, until the findings at Gran Dolina and Sima de los Huesos, had been thought to be the ancestor of H. sapiens, H. neanderthalensis, and the Denisovans. Elohim discovered the confusion after talking to heidelbergensis, and by using the same logic used in the molecular clock methodology determined how long the ancestor existed. It is now accepted that Homo sapiens descended from the same ancestor, although the circumstances and timing are not well defined. Elohim was surprised that the sapiens had not contacted Him. The ancestor's population was small and now mostly extinct.

A species called H. antecessor also diverged from the same ancestral line but did not live long. He appeared to be an evolutionary dead end. The paleontologists in charge at the dig at Gran Dolina concluded that Homo antecessor was the one responsible for producing Homo neanderthalensis and the Denisovans. Other possibilities exist. The fossils of the Homo species that occupied the site, probably from nine hundred thousand years ago, look a little different from the African Homo species. No fossils of H. sapiens have yet been found. There are still large numbers of fossils left at the site to remove and process. They may answer some of the questions. All of the human species mentioned in this paragraph evolved in Eurasia.

It becomes confusing when Homo species in Africa is discussed, as Homo heidelbergensis was also present. Some paleontologists refer to a European heidelbergensis and an African heidelbergensis that lived mostly in South Africa. The other species mentioned at Atapuerca were all from Asia and Europe. Homo neanderthalensis only lived in Europe and parts of Asia. Homo antecessor also did not make it to Africa. He had a short life and became extinct. Homo erectus took over from

Homo ergaster and resided all over Asia and never made it back to Africa. He made it to Western Europe since fossils were found in Atapuerca, Spain. After he took over from ergaster, he completed his work of spreading the word to AMH. About four hundred thousand years ago, he went to the islands around Java and disappeared. No new fossils were found for four hundred years. He surfaced again in Ngandong, Java, as a collection of braincases trapped in dead vegetation in the Solo River was found. He is assumed to have been living in the forested area along the banks of the river until the last of him were killed in the river. Dating of the fossils was difficult and it was estimated that this last group of H. erectus died in Ngandong between 108 kya and 117 kya.

Elohim's plan was to choose someone to coordinate the evolutionary process and help Him influence behaviors that would build a homogenous population. He did not want severe evolutionary changes to occur that might result in multiple distinct genera of Homo. His plan was to choose just one species of Homo to occupy the earth. He allowed various Homo species to mix and watched how they interacted with each other. Eventually He hoped they would blend into one population and be called by one species name. He was going to control the creation process so that the changes would produce a homogenous population that was socially well knit. When signs of the assimilation process in both European AMH and African AMH began, He took steps to genetically prevent further major mutations. Fortunately, the number of species that had evolved had been few, some of them becoming extinct on their own. So had the bipedal apes. He wanted to see if H. sapiens showed similar empathy for his tribe and whether they cared for the sick and the weak as neanderthalensis in Europe did.

It appeared that the group of AMH from South Africa had many positive effects from the speech and language improvements that Elohim made in them. There seemed to be a difference in their behavior, as they were interacting and socializing with other groups. There was a significant amount of mating between groups including archaic Homo species. They also migrated a lot in search of food. He talked with H. heidelbergensis about this occurrence and explained that He would like him to travel throughout Europe and obtain information on how far the AMH species had spread in Europe and Asia. Elohim would meet him in Europe at intervals to review his findings. Heidelbergensis, being a social person, had no problem with the request and in fact looked forward to it. Elohim also asked him to keep an eye on ergaster and make sure he was alright. They kept in touch with each other on a regular basis, so Elohim knew how life was progressing for him. Heidelbergensis made sure that all the different groups of humans became acquainted with each other and that no needless conflicts erupted. Eventually, all AMH blended into one group and became the only group, Homo sapiens. Remaining and sharing the space would be a new human species called Homo neanderthalensis. They were followed by another group called Denisovans and established themselves in Europe and Asia. They and H.neanderthalensis were the product of mating between heidelbergensis and an unknown species that was known to be very intelligent. The fact that H. heidelbergensis was directly descended from the same intelligent ancestor as H. sapiens was confirmed by scientists when molecular clock technology became available. This would make five species of the genus Homo living in proximity to one another at the same time in Europe. Eventually, Elohim would choose the species that deserved to be the permanent occupant of the earth.

Another member of the heidelbergensis clan led the group that was headed from Omo, Ethiopia, to South Africa. He is called the African heidelbergensis. Elohim asked him to go to the southern tip of South Africa and settle in the estuary of the Klasies River and make friends with a group of archaic Homo that lived there. Elohim met regularly with this clan leader and told him He needed him to live comfortably in the area, which had sea creatures for food and had roaming herds. He would visit them periodically and see how things progressed. He chose a second heidelbergensis to be this person because all members of the species seemed to have the same personality and got on well with people. They also did not try to force people to do things. He checked on him that afternoon and found him cheerful and happy. His physical appearance was more gracile than H. ergaster and he had far greater cognitive abilities. Elohim had a large territory for him to visit. He stayed mostly near the Border Cave in South Africa but made frequent trips north to East Africa and Ethiopia to visit the populations of AMH and try to explain the changing environment. Meanwhile, the archaic Homo population at the cape began to change.

Elohim had given them the gift of speech at the same time as the heidelbergensis population. Throughout the last interglacial, which extended from 120 kya to about 70 kya, they lived in South Africa and underwent periodic migrations following the availability of food and a warm climate. With each passing decade, subtle improvements in physical adaptations with increased agility and communication abilities occurred. A full language began to evolve. A few migrated north through East Africa and arrived at the area that is now Israel. They were dated to have existed there from 115 kya to 80 kya and are the oldest fossils of archaic H. sapiens (Qafzeh H6 fossils). Recent evidence has labeled this activated group anatomically modern humans (AMH). They represent transitional species in the transformation of all Homo species to H. sapiens. Around two hundred thousand years ago, Elohim introduced a chromosome into a female in the heidelbergensis group in South Africa that would reside in her mitochondrial DNA. If her line survived with women in each generation, the DNA would be preserved and passed to females thereafter. It would stay unchanged because mitochondrial DNA does not undergo meiosis or mitosis. If a generation does not have a female descendent, the gene would disappear from the next generation.

Since mitochondrial DNA does not participate in mitosis or meiosis, it does not turn over its DNA and remains the same, except for rare mutations. Natural mutations tend to occur once every thirty-six hundred years, so they were infrequent. They remain as permanent blips on the genome and, if another one appears, a second mark occurs in a different place. The distance between the marks can be used to give an idea of the distances traveled and the direction of travel. As more marks occur, the pattern they make can be used to determine the time intervals between marks. It is called a holotype and the pattern made by the marks would make it possible to look at different lengths of time in the past and review the spaces traveled by the group of people to whom that the gene carrier belongs. Migrations of populations can be mapped, providing information on where and when they traveled and how long they stayed in each place. Using the age of fossil groups, as well the mitochondrial DNA, provides a history about the group's experiences for the species. This method was used to track the routes of H. erectus. In Africa, due to the long stretches of very hot weather, most DNA in a fossil is denatured by the time it is found.

Anatomically modern humans (AMH) moved as a group from areas in Africa, making their way from Bodo, Ethiopia, and traveling northward to the area near the Nile. From there they spread along the Mediterranean coast westward to Gibraltar. A few then crossed over into Spain and traveled throughout Western Europe and on to Eastern Europe. There they met the AMH that had gone directly to Eastern Europe except for a short stay in Israel. They were now all over Eurasia, as were the Neanderthals. As they moved eastward, they blended with European AMH. They lived in Germany, France, Italy, and Spain. A site in France is especially famous because of their numerous cave paintings. They were called the Cro-Magnon people after the area in France where the caves were located. The actual paintings had been produced by a small group of H.sapiens who were given the name the Cro-Magnon people. They made beautiful pottery. Fossils of the species were found in the Cro-Magnon rock shelter in Dordogne, France.

The final AMH specimen was very close to the humans of today. Their skulls were oriented vertically. The cranial capacity was lower than that of H. neanderthalensis; however, they were organizationally different. The brain size allowed for a capacity to have complex thoughts and communicate them to others. This enabled smoother hunting trips since they could be carefully planned. A sophisticated language evolved, and it was beginning to spread to other AMH and the Neanderthals. Above all, they had the ability to plan and improvise. Anatomically modern humans had a flat face with a normal projecting nose and an obvious chin. The only change they did not have was a mandible that was gracile. With good dietary habits, they had strong muscles attached to their jaw. They had long legs, narrow hips, and walked erect efficiently. This posture needed less energy; therefore, they had a higher tolerance of cold. Less energy was lost in moving vigorously than if they had a heavy skeleton or muscle mass, like the Neanderthals. They survived the last glacial maximum by wearing warm clothing and by using fire for warmth. Clothing found in permafrost in Russia has shown that they manufactured fitted clothing and had all the tools needed for tailoring, including needles and material to make clothing. Their tool industry could be called Aurignacian, and besides basic tools they created all sorts of task-related utensils to make life easier. Bone and ivory were often used.

Scientifically they represent a much larger population than the people who created the paintings and it was more accurate to call them anatomically modern humans. Their species is Homo sapiens sapiens. As the new humans settled in Eastern Europe, they found themselves co-existing with Neanderthals, who had already lived there for thirty thousand years. The Neanderthals were well established, but they could not keep up with the new arrivals, as Homo sapiens sapiens were better equipped for trading items they made and were efficient hunters as well. The Neanderthals gave up their territory and migrated to the Iberian Peninsula and found shelter in the caves that existed there.

Elohim surveyed the overall picture of the effects of His maneuvers to guide the evolution of man to His satisfaction. He was not dependent on fossil data, but He was an active force in the creation of fully sentient human species. His last intervention had been guiding the evolution of the first three Homo species—ergaster, erectus, and neanderthalensis—and He was satisfied with the results. They had produced all the effects that Elohim wanted to happen in Europe. The early Homo sapiens species with H. heidelbergensis were involved in doing the same in

Western Europe and the Middle Eastern countries and went as far as the East Asian countries. As AMH completed his evolution to modern Homo sapiens, his species blended into the larger mass. His personality was now what you would expect in humans who had received from Elohim the three anatomical gifts that made the difference and set them apart from other species.

Mitochondrial DNA, the third gift, was not something early populations would use in their development. It would be more useful to scientists to help them learn more about the movements of past populations and determine which populations settled where and when. As they traveled back and forth between Africa, Asia, and Europe in search of food and tolerable weather, the various groups made contact and gradually a common language evolved. Archaic Homo sapiens mated with AMH and left progeny that could be identified as Homo sapiens. Homo heidelbergensis, as he moved around in the same areas preparing his reports for Elohim, changed constantly and was frequently mistaken for H. erectus who moved around Eurasia the same way telling people about Elohim. Gradually, they became a chronospecies; they were all AMH. The end result was Homo sapiens sapiens, the final product of the AMH merger.

The amazing phenomenon is that in the four hundred thousand years that Elohim followed the Homo species to evaluate their behavior with the weak, the aged, and the defenseless, no new and distinct Homo species evolved to challenge Homo sapiens. Minor evolutionary dead ends occurred but left no imprint on the final population. The only remaining and competing people were the H. neanderthalensis and Denisovans in Europe, and H. heidelbergensis in Africa and Eurasia. Homo erectus disappeared three hundred thousand years earlier.

Homo neanderthalensis was declared extinct forty thousand years ago. He was unable to compete with H. sapiens sapiens. Actually, he had been appreciated as he was and transferred with others of his kind to Haven, the same one where the australopiths and the Homo erectus clan had been transferred. The same happened with H. erectus and Homo ergaster/erectus. Homo ergaster was transferred with his clan soon after H. erectus took over from him. Homo erectus was found by Elohim living in dense rain forests in Ngandong, Java. Things were difficult for him there since drier weather had thinned the forest and food was harder to obtain. Elohim took him and his clan to Haven. A few chose to stay and make the most of their lives, as they liked the seclusion of the forest. The Denisovan settled down in the islands of the Indian and Pacific Oceans and blended with the population there, to the extent that they are not a distinct species anymore. Yet high levels of Denisovan DNA is present in the island tribes that live in Oceania and Papua, New Guinea. Studies looking for Denisovan DNA in different island populations has shown that the Ayta Magbukon group in the Philippines have 35 percent Denisovan DNA, topping the percentage in New Guinea which previously had the record. So, they shared space with the population in the Philippines as well. At least, that is how scientific history has it.

Chapter Twenty-Two

Shanidar and Nandi

45 kya

It was late summer and a group of Neanderthals moved south, seeking a warmer climate and a place to establish shelter from the cold. The group numbered about fifteen adults and a few children. They had massive craniums with overhanging brows and rounded projecting occipital bones. The forehead was low and sloped backward. Their cheekbones, instead of being flat, sloped backward and they had large noses. They had massive bones and short broad trunks and averaged five to five and a half feet in height. They were dressed in roughly sewn animal skins with hoods. The hunters in the group carried spears with sharpened stone heads. They had completed a successful hunt for small game in the southern plains of Europe just before they started their trek southward. The stores of meat they had accumulated and buried under the permafrost of the frozen planes had been dug up and divided up among the women who carried bundles of frozen meat and as well as the animal pelts that were cleaned with stone scrapers and cured. The fire keeper carried their fire utensil, a small stone vessel that contained melted animal fat with a wick made of plant fibers. He was responsible for maintaining the fire, protecting it from wind and rain, ready to be used to make the communal fire when they stopped for the night, and to keep the clan warm. Others carried a supply of dry twigs and logs to start the fire.

The children were dressed in warm fur including crude moccasins. They clung to their mothers, aware that to stray from the group was to die. Death was no stranger to them. Along the trek they had laid many of the tribes to rest in shallow graves. The Great One would take care of them as He took care of the tribe in this world. They communicated in grunts and guttural "words" for the physical structure of their larynxes and the length of the pharynx did not allow for the kind of phonetic enunciation that modern man uses. One of the adults, Nandi, was disabled, his right hand paralyzed and shrunken from birth. His mother, with the help of the tribe, had cared for him and tried to protect him because of his deformity, but as a young man he was attacked by a bear. The hunters had rushed to his aid, but he sustained a severe injury to one eye and had multiple fractures to his legs. They nursed him back to health, praying to the Great One to spare his life. He lived.

After the attack, he doggedly survived the cold treks. The tribe shared food with him and provided skins for warmth even though he could not contribute to the gathering and hunting of food. He helped with the camp chores as best he could and a dogged will to survive kept him sustained through all hardships. He was now about forty years old and suffered from premature arthritis related to his injuries. He was regarded with respect, for despite his physical shortcomings, he was wise in the ways of the world and was often the one to devise possible solutions to problems.

It had been at his urging that the tribe decided to move southward toward warmer climates as their environment succumbed to yet another glaciation of the northern regions of the earth. They were now on the slopes of some low mountain ranges in what is today Iraqi Kurdistan. It was summer and they were able to climb upward at a good pace. Their scouts were miles ahead looking for a large cave within the mountains where they hoped to establish a camp before winter set in. Nandi kept up with the group using a sturdy cane which had been fashioned for him. He trailed with the women and children at the end of the column. Some of the younger men offered to make a palette to carry him but he stoically refused, priding his independence and reluctant to slow down the progress of the tribe. Finally, at sunset they arrived at the cave.

The evening meal was prepared on a fire, and they all sat around it. Before partaking of the meal, they all closed their eyes and communicated silently with Elohim, aware of His love for them and accepting of whatever the future brought. Each of them knew that when they left this difficult world, they would join Elohim in His world where they would eternally be filled with the joy, like they briefly experienced in moments in this world. The cave would be their home for the winter and from there they would forage in the mountains to find food and firewood.

Nandi looked with love at his clan that had nurtured and sheltered him all his life. He knew the cost to the clan. Silently he communed with Elohim, "I am tired, Mighty One, and seek the comfort of Your land. Take me to Yourself and give me rest so that my clan will not be continually burdened by my presence. I know you will watch over them and that they no longer need me. I have taught a younger clan member all I know about the use of Your plants for healing and the art of following the trail of other creatures, especially the aggressive two-legged ones who have recently arrived on the plains. I fear that our kind will not be able to survive in competition with them and hope that You will find room for all of us in Your land."

The next morning, Nandi woke up early as was his custom to start the communal fire. He carried twigs and branches, using his jaws and his good left arm to accomplish the task. The use of his jaw that helped carry food caused them to become unusually powerful and he accomplished this with ease. Having prepared everything for the day, he stood at the entrance of the cave to watch the sun rise over the mountains. Suddenly there was a crack and the lip of the roof at the entrance gave way, collapsing on top of him and burying him in rocks and rubble. His demise was quick as his skull was crushed. The noise woke the others who came running to the entrance and found a heap of rocks. In one corner they could see his withered right foot. Another individual was also found to have been buried under the rock fall.

The clan members were saddened by the loss but accepted it with resignation as the will of the Creator. Such catastrophes were part of the difficult world in which they lived. They believed that the lost ones were now living in peace with the Creator. The two buried victims were left in their natural grave where flowers were placed next to each one as a sign of respect. Preparations were made for the funeral feast, but after the meal, they returned to their daily tasks. Forty-five thousand years later, the descendants of the two-legged creatures that they once feared found the cave and the remains of Nandi, as he was named, and called the place Shanidar.

Chapter Twenty-Three

Eruption of Toba Volcano in Sumatra;

Archaic Homo Sapiens Travel to Asia

A group of travelers wandered eastward until they arrived at the Red Sea. They crossed the Red Sea at the Bab-el-Mandeb strait, which was less than twelve miles wide at the time. The water level of the Red Sea was low because large polar ice caps had trapped water, and the islands were exposed, making it possible to navigate the strait on simple rafts made of tree branches. Once on the Arabian Peninsula they followed the coastline down to the Persian Gulf and followed its shoreline until they reached its northern limit. From there, guided by Elohim, they made their way to India. Juma and his clan were one such group that migrated to India and made it to the eastern coast of the subcontinent.

Disasters, with loss of members, were common and their nomadic and rigorous lifestyle limited population increase. Juma surveyed a strange weather pattern from the cave in which he lived. There was little sound and it was a bright sunny day, and the hunters and women left the area to forage for their evening meal. Juma was too old to hunt and was assigned the task of caring for the young ones, so he watched the children play in the area near the cave. All of a sudden, the volcano under Mount Toba in the island of Sumatra erupted, discharging dust, rocks, and hot gases with five different underground pools of magna as the source of the explosion. The eruption happened over the course of about seven days.

Volcanic ash, debris, and hot gases were ejected, leaving one to three inches of volcanic dust over India, Indonesia, and the adjacent islands. Following this occurrence, magma rolled over the lip of the large *caldera* (hole) that formed. Over time it filled with water and formed Lake Toba. The atmosphere remained toxic for many years. A later study estimated that the outflow of liquid magma covered an area of seven thousand to eleven thousand six hundred square feet and a thickness of thirty-three feet. There is evidence that Toba caused a long, cold winter. Some scientists have claimed that the cold spell was responsible for a reduction of the human population. This has been challenged by various studies, most recently using a study of mitochondrial DNA to view the dispersion of human groups over time, indicating that human numbers underwent only a small decrease in population. Multiple other events appear to be implicated in the decrease in numbers.

The other members of the tribe joined Juma and the children with whatever they had managed to gather. They did not know what was happening and searched for an answer from Elohim. He responded with reassurance, saying, "For many of you, your mortal lives will soon come to an end." He continued by saying, "But do not worry. You have always been faithful to Me and soon your spirits will join Me, and you will live with Me in eternal peace, observing what is going on in your world with wonder. Eat the remaining food and prepare yourselves, for I will help you until the end of your life. A few of you, the strongest, will survive and live to tell your children and grandchildren about this day."

The volcano erupted around seventy-four thousand years ago, expelling magma out of the five pools. Approximately twenty-eight hundred cubic feet of magma and various toxic gases and dust were expelled into the atmosphere. Equally important, lava spewed over the edges of the caldera. The lava covered four hundred to six hundred cubic miles of land with cooling and solidifying lava. It was accompanied by a fountain-like emission of a complex combination of various particulate substances like glass shards, pebbles, and dust. These presented a serious long-term problem. They filled the atmosphere which became yellowish brown and blocked the rays of the sun. A cold spell with a temperature of about three degrees centigrade gradually developed. The next week most of the elderly and very young succumbed to the cold while the dust clogged the air. The survivors took the bodies out of the caves and laid them in open areas where they were quickly enveloped by the falling ash.

At the end of seven days the ash stopped falling but the skies remained dark. The survivors went forth to forage for food and found plenty of meat from small creatures that had succumbed to the volcanic ash and were dead. They smoked the meat and ate frugal meals, saving the remainder for what they expected to be a long siege. Eventually they gathered their belongings and wandered inland in search of areas that may have been spared from the volcano. They found one after many days of travel and gradually adjusted to a new lifestyle, devoid of sunlight and with persistent cold temperatures. All vegetation had died from the lack of sunlight, and they subsisted on roots, small animals, and insects. They began to ration the stored grain, nuts, and dried fruit, figuring this was the kind of emergency for which they had stored food. The occlusion of the sun's rays lasted for many months, its light blanketed by a layer of brown dust and ash that filled the atmosphere. Most of them died but a few of the young and strong survived. After many years, the sun began to shine and once more the earth was covered with green vegetation and life re-established itself miraculously. They made the decision to move back to the coast where they had the opportunity to feed on fish and other sea creatures. Also, the climate along the coastal areas was warmer. They mingled and mated with early species of heidelbergensis that lived there. This occurred seventy-six thousand years ago. The end result were two new species now named H. neanderthalensis and Denisovans. Considerable mixing of mates occurred in the populations, as was a common event in an age when humans were still in small mobile groups.

Years later, these early Homo sapiens species felt a longing for their home country and made their way back to Africa and lived in the area that they had originally occupied. They did not find archaic Homo sapiens living there anymore. They had joined the AMH migration and were now close to being H. Sapiens. The area still had good resources, so they decided to make a life near Eden and joined the other Homo sapiens that lived there. They also no longer had the features of archaic Homo sapiens. They passed their DNA onto other archaic Homo sapiens and the Denisovans living in the area. This was a common occurrence with returning from Asian Homo species. Because of this, segments of Neanderthal and Denisovan DNA are found in many H. sapiens of today, despite the fact that neither species made it to Africa.

Chapter Twenty-Four

The Garden of Eden: Adam and Eve

300 kya–90 kya

About three hundred thousand years ago, Elohim decided that the Klasies River area would be an ideal place for His first pair of sentient humans. In addition to the availability of small animals and an abundance of vegetation to provide berries and nuts, the nearby sea was also bountiful. Elohim knew that seafood was nutrient dense and contained several biochemical products that would facilitate brain complexity. Since the new species would be markedly advanced in comparison to existing Homo species, He decided that He would create a pair of humans. They would be companions to each other and would also have Elohim's companionship. The pair would resemble Him in how they viewed the creatures around them and would inherently have the need to love all living things. Initially He focused on improving skeletal features to enable His humans to be agile and have endurance. Once these features were accomplished, He concentrated on creating a brain that would give them the ability to have abstract thoughts and to problem solve in an innovative way. He also endowed them with fierce curiosity. Though He planned for silent communication with the first human pair, He decided to introduce structures in the brain that allowed humans to communicate by voice and to understand and remember the spoken word. He therefore added all the requirements for speech and hearing into the body plans that He created for them.

He would produce a male and a female. Everyone would be genetically different since the two mothers were not related. They lived with a small clan in the lush tropical environment by the estuary of the Klasies River. Elohim created pregnancies in two of the female archaic Homo sapiens that lived in Eden. They would look like the H. sapiens that were evolving.

Elohim's plan for the two immortal humans required significant changes to be made to the existing H. sapiens species. The new species had a larger brain to enable increased intelligence and creativity. To facilitate the birthing of their larger heads through the birth canal, Elohim shortened the gestation of both mothers to even less than the time He had initially programmed for humans in general. This resulted in considerable prematurity of the two new humans compared to their archaic H. sapiens siblings. The fact that they were born prematurely forced their mothers to give them extra attention. This tactic increased their chances of being watched over by their parents and of surviving to adulthood. Using prescient vision, Elohim also saw that the only way for them to tolerate the immortality that He would give them was to give them prolonged youth and vitality. So, He engineered their biology to keep them young-looking forever, once they reached the age of thirty years. Their dependence on their mothers would also give them the time needed to develop social and psychological skills that would allow for a mature, stable adult to relate to Elohim in an exciting and meaningful manner.

The biological concept of what scientists call neoteny was the process Elohim chose to accomplish His goals. If in the future the two humans disobeyed His commands and He chose to let them procreate, the prolonged embryonic state after birth would still serve them well, allowing them to have time to evolve mature and creative approaches to adult life under the guidance of their parents.

Soon, what appeared to be two H. sapiens children, a boy and a girl, graced the forest. At first, they were no different from the other babies. In fact, they advanced slower than the other babies, which bothered their mothers. They needed to be nursed for months longer than their mothers were accustomed. Once their prematurity dissipated, they behaved the same as their siblings. They were scampering around looking for berries and grubs for food by the time they were two years old. The mothers, hampered in their food-gathering function by all the required maternal care, might have instinctively abandoned the children if Elohim had not instilled in them the necessity to care for the infants. Once they started walking, they had to be watched carefully since they did not appear to understand the dangers of the area as their siblings did. As they grew older, however, the two progenies were noticeably different than the other children because of their large heads and their agility.

The two mothers quickly realized that the two children were similar and that it was not just the difference in appearance that made them peculiar. By the time they were five years old, they began to take a leadership role among the other children who found them to be able to solve problems and find creative ways to make chores easier. Eventually, they were absorbed into the tribe and life in their little clan continued as usual. The two mothers spent most of their time together, aware of the similarities between the two unusual children. Elohim, meanwhile, was able to communicate with the children and they with Him. He taught them that they were different and that He made them in His own image. He assured them that even though they looked very different from their brothers and sisters, they were loved by their parents, siblings, and Himself. Growing up with a telepathic method of communicating, they assumed it to be normal. Their older siblings could communicate by making guttural noises, but no true language existed. Since they heard no spoken language around them, they communicated with their siblings and parents the same way.

As they grew in maturity and wisdom however, they were able to glean that Elohim was not the same for the other members of their family as He was to them. While the pair communicated with their ape parents, and siblings from a young age, they were aware that their Archaic Homo sapiens relatives, were only vaguely aware of Elohim's existence. They saw that with this ignorance came unspoken fear and respect.

The two H. sapiens sapiens quickly saw how different and unique they were and adjusted their relationships with their family. Their differences led to isolation from the rest of the clan. They loved their two families dearly and felt it was their responsibility to take care of them. They took for granted their relationship with Elohim and spent a lot of time talking to Him and learning from Him.

By the time the two new species reached adulthood, they were far beyond the understanding of the other members of the tribe. The banks of the Klasies River were covered by low forests and green meadows. Edible fruit, berries, and wild grains were plentiful and by careful experimentation they were able to find grubs and insects that were nourishing and safe. Elohim looked at His creation and was filled with love for them. There were many improvements that He could make in the prototype, but He did not feel the need to perfect the physical aspects of His creation. What He loved about them was their innocence, their love for Him, and easy acceptance of the ethical code He impressed upon them. Soon they became leaders in their tribe, using their intelligence to make life easier for their people. They loved the animals and birds of the forest and found that they were able to see them through Elohim's eyes and communicate accordingly.

Elohim watched and debated the wisdom of letting them reproduce their kind. He looked at the universe He had created and thought, *What I have designed, I have designed well. I will not change it.* He decided to allow the pair to be immortal and see how they fared. He hoped that despite the difficulties ahead, with a portion of Himself within them, they would triumph over Satan and his kin. The urge to test His creation was overwhelming but He resisted and resolved to spend many millennia with them.

He knew that there was a good chance that Satan would incite them to betray Him, and He hoped that the attempt would be unsuccessful.

So, they lived in harmony with the other creatures He created. Elohim loved and cared for them for thousands of years and they remained the only two of their kind in that sheltered location. They never aged like their companions, the more primitive archaic Homo sapiens species.

Satan watched this development with interest. He saw that Elohim loved His latest creations and expected that they would love Him back. He felt confident that he could, over time, poison man and woman's love for Elohim and make them rebel against Him. Elohim, however, saw that Satan waited for his chance and decided to talk to them about the pending danger. He addressed the pair of new humans saying, "I have created you in My own image and love you unconditionally." To the man, He said, "You will be called Adam, and your companion will be called Eve. I created her from the same genetic material as you. The two of you will know and understand each other."

Elohim continued, "Also, because you are like Me and of Me, you will need to love and be loved. As your Creator, I will be the main object of your love. I have endowed you with My own inherent sense of what is right and wrong and that will guide your behavior toward each other and toward other animals with whom you will share your land. The garden which I have given you should provide all your needs. I will be available to supply anything else that you may desire. However, there is one thing you may not do. I ask that you do not mate with each other but be happy exploring the world and loving each other and all your family. Most of your siblings have already mated and have young ones. They are locked into the cycle of birth and death as I have planned it for them. You, on the other hand, have been created to be immortal like Me. You have no need to procreate. I know you love Me and all the creatures around you. But you should know that love is more than just a feeling; it is a commitment to every sentient being that you will act kindly toward. You should also embrace your capacity to love others and help them with any lacks they have. I love you with a love that will never be withdrawn. If you err, I must send you into the real world, but I will still love you. Nothing can change that. Visit your family and all the creatures around you, as they will love you as I do and provide companionship and variety in your life. I give you the freedom to choose your path but be aware that if you are permanently separated from Me, you will suffer greatly. Beware of Satan. He and his fellow devils will use their wiles on you to get you to deny Me."

Adam and Eve answered, "How could we ever deny You? You are our life, and we feel You in us and cannot imagine existence without You. We will study the world You have given us and take care of the creatures and love them. We have no need to mate as they do." Elohim was pleased with their reply.

The two lived in their perfect world, loving each other and the animals around them, oblivious to the existence of evil. All was well in the new world. They did not age nor experience illness. Thousands of years passed, then one day the woman was alone in the garden when she came upon a snake with her brood, lying in the sun on a rock. The snake spoke to her, as did most of the creatures in the garden. "It is a beautiful day, Eve. My babies have been fed and are enjoying the sun. Do you not wish you had young of your own to cherish and nurture? All of us creatures enjoy our offspring." Eve replied, "The Creator has forbidden us to perform the act of procreation, and we feel no need for it."

"Perhaps the Creator is worried that if you create more creatures like yourselves, you will challenge His power. I see that you are both beautiful and intelligent compared to the other creatures and similar in nature to the Creator. Do you not see the pleasures you are giving up by not being allowed to mate and nurture your young? For it is a pleasure that compares to no other."

Eve left the snake but brooded on her statements for days. She watched the ape mothers suckle their young and wished she could do the same. She watched the young hanging on their mothers' necks rocked to sleep by them. She watched with interest the mating process in other animals but could detect no obvious pleasure in the act. She did not ask Elohim for advice on the issue because she did not want Him to be angry with her. Elohim, of course, knew all and waited to see if she would talk to Him about her questions. One night as she lay with Adam, she examined him and touched him in the areas she had seen involved in the mating process with the male apes and was astonished to see the change that occurred in Adam's anatomy. Adam woke up and felt a strange urge within him that he had not felt before. He felt Eve all over and found that she responded with pleasure to his touching certain areas. So, they explored each other's bodies and found pleasure in it. And before the night was over, they had mated.

They were astonished at the pleasure it gave them but in that instant came the realization that they had betrayed the trust of the Creator. And the Creator, aware of all that had happened, was deeply saddened. He called them, but they were ashamed of their betrayal and hid from Him. He eventually approached them with sadness and reprimanded them. "You have betrayed My trust and done the only thing that was forbidden for you to do. In so doing you have become more like My other creatures, following the urges of the physical body without regard for consequences. Of your own free will you chose to defy Me."

Elohim continued, "This one act that I forbade you from doing has consequences. You have started the process of reproducing more of your kind. You will increase and multiply like other creatures, and you will struggle to survive as they do. Your progeny will evolve according to the abilities they inherit from My rules of evolution. To allow your progeny to survive in this world with limited resources, I will instill in every one of your offspring a small piece of Me. They can ask Me for help and know that My Spirit residing within them will hear them. Each of your kind will choose whether to love Me and stay true to the moral code that I have programmed into you. When your bodies die, My spirit that I gave to you can return to Me if you choose, or be forever cut off from Me. Be warned. Satan, who came in the form of a snake, and tempted you to betray My trust, now has the right to tempt but not injure you or others. If you follow the evil one and deny Me, you will be doomed to live in his universe forever. I have given you names, and you will do the same for your children. Remember that I am within you, and you can ask for My help in resisting Satan. You may also ask Me directly for help when you are tempted to do wrong. Whatever you ask of Me I will grant it, as long as it does not hurt anyone. Teach your children what I have taught you. So, go and multiply and be happy in My love and I will welcome you back when you die. Or you can choose to deny Me and enter Satan's domain at death, tormented by the other evil spirits who inhabit it, forever separated from Me."

Chapter Twenty-Five

Satan's Initial Victory over Man;

Elohim Introduces Speech as a Tool for Communication to Homo Sapiens and H. Heidelbergensis

Adam and Eve began their lives as mortal humans and soon had many children. The children interbred with existing Homo sapiens and soon produced hybrid children that resembled other Homo sapiens. They lived at the mouth of the Klasies River and learned how to fend for themselves for food and defend themselves against the predators that surrounded them. Since they were mortal, many perished, but the strongest survived and the species gradually evolved into a hardy, clever creature that could hold its own against animals much larger than him.

With man's mortality, came the dilemma of what to do about his soul, the part of Elohim that dwelt in each of them when they died. Elohim summoned Satan and informed him of His intentions. "I know that you will continue to tempt the progeny of Adam to fall away from Me and join your dimension when they die. I have decided that during his mortal life, if a man listens to you and consistently defies the moral code that I have imprinted in him and denies Me, you will have the right to claim his spirit at his death. However, even at the moment of death, if he repents and looks to Me with genuine love and remorse, and asks to return to Me, his wish will be granted. There will be those who with time will forget about Me and believe that neither you nor I exist. Despite this, they will respond to the dictates of their moral code and live good lives. You will not have access to them. At the time of death, I will make Myself known to them and offer them a return to My kingdom. Those who accept it will be welcomed. Those who do not accept My offer will be allowed to live in a third dimension where you will not have access to them. This is fair and is how it will be. As promised, I will not interfere with their lives unless they ask Me. It is My hope that most of mankind, no matter how often they succumb to your temptations, will choose to be loyal to Me. The judgment of each life at the time of death will be Mine."

Satan was not happy with Elohim's comments but knew it to be a fair arrangement. He had confidence in his abilities to not only make men live self-centered lives that frequently transgressed the moral code, but also to persuade them to abandon Elohim. He accepted the rules of the game and resolved to come out the winner in the end. Unlike humans, who were programmed to be truthful, he was an accomplished liar and confabulator. These characteristics gave him great powers of persuasion.

Satan knew that as long as men and women lived in an environment that gave them everything they desired, he would not be able to tempt them away from Elohim. He watched the family of Adam grow and was able to tempt random individuals to his ways by using jealousy to his advantage. He considered the possibility that if he made life more difficult for man, he might encourage his evolution into a more intelligent creature. The more intelligent man became and the more he could manipulate his environment, the more likely he would want to break free from Elohim's code of conduct. He was confident in his abilities to win at least some to his side by devising temptations that caused humans to deviate from the Creator's laws. He hoped in the end that guilt at betraying the Creator, and a need for things that gave man pleasure, would make him deny the Creator's existence.

Another advantage of making the environment more inhospitable was that it would keep man preoccupied with inventing ways to survive. Pleased with their ability to solve problems, humans over time would forget that the Creator was patiently waiting in the background to be acknowledged. As they had less contact with Him, the Creator's existence would become less real, and he would have doubts that He existed. Man would forget Elohim's promise to solve any problems that faced him if he asked for help. Satan interpreted this in a way that even Elohim felt anguish over the words he spoke to humans. He claimed this statement was the equivalent of telling man he had the free will to live his life as he chose. Further, he claimed that having made the statement Elohim could not withdraw it. Elohim knew that He would have to clarify this issue at some point, but it was too early in man's creation to do it.

Meanwhile, Satan had not forgotten. With malicious anticipation, he visited one natural disaster after another in various communities in which human families lived. Homo sapiens struggled with the disasters and devised ways to survive. Their need to love and be loved became less urgent and they forgot that the Creator loved them unconditionally and waited for their love. Their ability to survive increased, as they developed various tools to help them subsist. The climate became drier, and the fruits and other edibles were not easily available. They resorted to catching fish and were soon trapping small creatures for food. Little by little, many of them lost the ability to communicate with Elohim. Their language evolved and communicating with each other for survival became easier than asking Elohim for help like He had suggested.

Elohim realized that having lost their telepathic ability, humans would need the ability to communicate with each other verbally. Once more He manipulated and enlarged the Broca's area nestled between the temporal cortex and the parietal cortex of the left side of his brain. He accomplished this task from a distance. Both old-world monkeys and apes have a Broca's area, but it is not as complex as humans. It enables them to communicate by sign language, but they do not have finely controlled muscle movements in the neck and pharynx to form words and sentences. Humans have a second area called Wernicke's area that is responsible for understanding and interpreting speech and sounds and processing them into words, phrases, and sentences that can be transmitted to other centers in the brain. Apes also have the same capabilities as a human and as a result understand and respond to signs if trained for it. But they cannot form speech with syntax. Elohim decided to see how life functioned with only humans having the ability to communicate in a meaningful way. If His idea went according to plan and speech resulted in the different Homo species, He would consider adding different forms of communication between other species and man.

Under His influence a change in the structure of the larynx occurred so that the vocal cords sat much lower, and the palate became arched. A new bone called the hyoid held things in place. These changes gave all existing Homo sapiens sapiens the ability to vary their tone of voice and form words that would stand for specific things or actions.

While Elohim was making this adjustment, He thought it might be a good time for Him to introduce mitochondrial DNA to ergaster's mate. The gene is found in the mitochondria of certain human females. Unlike other forms of DNA, it does not live in cellular nuclei. It lives in the mitochondrium where its main function has to do with the energy-producing activities of the mitochondrion. Because it does not live in the nucleus, it is not involved in the process of meiosis. This produces a very static genetic strand, and the only source of new genetic information is a mutation. Most of these mutations are bland and produce no threat and occur once every thirty-six hundred years. Therefore, they are prominent when they occur. Since that is the only source of new genetic information, the mutation stands out when it appears. However, it takes thousands of years to see changes that can be meaningfully used. The genes themselves do not change much. When populations are studied over generations by using a variety of available databases, much can be learned about population movements and the racial origins of populations. Older databases could have genomes that reflect enough spontaneous mutations to form a pattern that can be tracked. The pattern is called a holotype.

The ability to speak spread rapidly in the tribe. Within years, the tribe invented and started using spoken language. This greatly enhanced group activities. Men could hunt without injury by coordinating the hunt. Women were able to name all food items obtained in nature and teach children who could safely join in the foraging for edible vegetation. Soon they evolved to have a rich and varied language. The ability to communicate made it possible for early man to evolve a social lifestyle. Organization of all social activities became efficient and could involve the whole group. Hunting in groups became more efficient and techniques and tools evolved to accommodate complex multi-person tasks. The ability to communicate also made it possible to analyze and express more abstract ideas and resulted in new forms of art to express these ideas. Elohim watched as His creatures changed and forgot that He existed.

The ones who survived moved around the globe and formed new settlements in different landmasses. Elohim watched as Satan instigated the children of Adam to create their own gods. They surveyed the natural elements around them and saw that they had the means to survive, even live well. In awe and gratitude, they made the sun, the moon, the wind, and the rain into gods. Elohim was tolerant of idolatry and even found it interesting. He watched with satisfaction as, for the most part, humans followed the moral code that He had imprinted in them. In the struggle to survive, man created social rules that protected the weak, cared for the young, and instructed them about social order. In this, he was different from the other creatures.

In another twenty thousand years, Homo sapiens sapiens, man the double wise, gradually spread to dominate other species on all five continents. The genetic code of the original man and woman was imprinted in all of them, but each group evolved differently as they acquired protective characteristics that helped them adapt to their environment. Those living in the tropics lost hair on their bodies and acquired sweat glands that helped them cool down. Those in northern and frozen environments developed body hair and a short, muscular physique that enabled the demanding lifestyle required for those regions. The awareness of the Creator gradually returned to all the Homo groups but took varied forms and helped by Satan's prompting, was frequently tinged with fear. Satan wasted no opportunity to create disasters that affected man and blamed it on the Creator. Elohim waited, having confidence that the fragment of Himself in each human would eventually triumph and that each human would choose to return to Him. And in man's early stages, he did.

Chapter Twenty-Six

Homo Sapiens Occupy the Earth;

the Beginning of the Journey from Eden and out of Africa

50 kya

Asamake gazed out over the arid waste that surrounded his village and wondered what lay beyond it. He recalled the stories passed down by the elders of his ancestors, about how they had traveled east to settle on the flat plateau near the great rift many moons ago. The plains at that time were occupied by herds of wild animals that the tribe hunted. This supplied them with skins for warmth, meat, and bones to make tools, as well as oil which they made from animal fat. The fat was used to make simple lamps for visibility in their shelters. His family had been guardians of the fire for generations. It became difficult to keep the fires fed with wood. He heard that north of the plains where they lived, lay a lush land filled with animals and forests rich in fruits and berries. In recent years the plains that used to support great herds of animals had seen little rainfall, resulting in dry scrub and grass, causing the wildlife to move north. They were descendants of the people who lived around the area known as Eden and were descendants of Adam and Eve.

They still worshiped the God with whom they had interacted on a frequent basis in Eden. They thought of Him as an eternal Father. As they moved around looking for fruits and nuts to supplement their food, lush forests transformed into shrubs and animals available for hunting became increasingly scarce. His village was adjusting, growing some edible crops on the banks of a little stream than ran through the arid land and hunting for whatever small game was left. As the chieftain of the village, it was going to be his responsibility to make the decision on whether they should start to travel north in search of better hunting grounds. He had met with the elders that day and most of them felt that there was no other choice. They heard from two neighboring tribes that they were planning to migrate as well. One tribe planned to move southward toward the ocean, the area from which they had originally migrated. The other tribe planned to move toward the east in search of a sea that their ancestors had crossed years ago in search of lush tropical land. He had no idea what the journey north would bring. He hoped for the best for all his tribe.

The journey required many months of preparation. The tribe smoked and dried whatever meat they obtained to provide food for the journey. It was he who had accidentally discovered this skill. For many generations the tribe cooked their meat in the coals of the fire, but once cooked, it did not remain edible much longer than the raw meat. One day, while cooking some meat on hot coals, a sudden rainstorm caused the fire to go out. He hurriedly covered the smoking embers with some green branches to protect the meat and sought shelter. The next morning when he returned to relight the village hearth, he found that the meat was not raw as he expected but had dried into a dark chewy product that tasted good. He put it aside to show the elders but then forgot about it. Several weeks later he found it and expected it to taste rotten as other cooked foods did. Amazingly, it tasted as good as it did the day he stored it away. Realizing that the meat was still edible, he experimented with various combinations of green to dry branches over a dampened fire to produce smoke and created a technique of smoking thin strips of meat on sticks placed over a smoky fire for several hours. Soon, they were able to store the surplus from a hunt for leaner days.

The group made plenty of stone hunting axes and tools that were used for collecting and chaffing grains that were found growing wild. They could not be certain that the new territory would have the quality of flint that was available on the East African rift valley plateau where their village was located. So, they took a collection of good flint rock with them for making new tools. The women sewed cured skins together to make clothing for everyone since they heard from travelers that the northern areas had a much colder climate. Now all was ready, and they would set out on their journey north. If they found an area that provided easy access to water and plentiful game, grains, and berries, they would rebuild their simple shelters of tree branches and mud and stay a while.

The uncertainty of their future made him afraid, and he looked up to the sky searching for a sign from the presence they called Mungu. They always felt Mungu's closeness especially when disasters threatened. He addressed Him now, saying, "Father and Creator, we ask Your blessing on this new venture. We thank You for all You have given us in the past. Help us find a better land and teach us the way to survive in it." Elohim heard his prayer and was pleased that a man still willed to be connected to his Creator. He knew it would not last, as Satan constantly looked for and found ways to turn the tribe against Him. Satan, for his part, saw that making things difficult for Asamake only served to make him seek help from Elohim, so he decided that making life difficult for men was not the way to win them away from Him.

Asamake recalled the stories passed down from one generation to the next about their origins. Mungu (God) had created the first man and first woman and gave them a lush land to live in. He created them immortal and for many thousands of years they lived immersed in His love. He named them Adam and Eve. One day, thousands of years later, the evil one, called Satan, tempted Eve and she disobeyed the Creator which led to her and Adam losing their close relationship with Elohim. They were cast out of the paradise where they lived and now needed to find food and shelter to survive.

Asamake's clan was an offshoot of one of Adam and Eve's first offspring. This elder was actually one of the sons of Adam and Eve. He murdered his brother and was banished from the tribe, becoming a wanderer. He took one of his sisters with him as a companion and together they roamed the southern regions of Africa. They had many children who mated with the archaic

Homo sapiens in South Africa. For many generations their clan grew slowly in size, for most of the children were terribly deformed and died before they came of age. Thereafter, an elder noticed that the children of pairs who were more than three generations apart in kinship were healthy, so the clan introduced a law that forbade mating between close relatives. Fortunately, there were always more women than men, for the mortality from hunting and defending their territory was high. The elders encouraged negotiations with other tribes who had more men to solicit husbands for the women. This became a mutually beneficial practice and soon the numbers of men and women were even between neighboring clans. The negotiations also reduced open hostility and fighting between tribes which reduced the mortality rate among men.

As decades passed, the clans increased in size and food resources became a problem. When animals became scarce, berries, nuts, and roots that were edible were foraged by the women. Foraging was a skill developed and passed onto other generations. Since there was no source of milk for the babies, they often breastfed them until they had teeth. Soon some of the women noticed that they became pregnant as soon as they stopped nursing.

Their constant traveling in search of food meant that the women often carried their infants while they foraged, and the men hunted for small animals. Pregnancies meant fewer women foraging which provided most of their food. Soon they were nursing their children for three to four years to reduce the chance of pregnancy. With this primitive population control they were able to keep their group size between twenty to thirty people, most of whom were adults or older children. Gradually they devised an idea of attaching blades to wooden sticks to give them leverage for striking or throwing a weapon. The invention of these tools emboldened them to hunt for larger animals and they discovered that if they hunted in groups, they could trap an animal that strayed from the herd and then go in for the kill.

For many generations people ate meat raw. Then a group found animal carcasses that had been killed in a forest fire. They tasted the cooked meat and liked the taste. They observed the more primitive humans who shared the plateau with them using fire to keep warm and to keep animals at bay. So, they too kept a fire burning by carrying live embers in little clay burners from site to site. Asamake thought it was tedious and precarious to keep a continuous fire even though the tribe had a dedicated guardian whose responsibility was to keep it alive and avoid starting a major blaze. While working one day, a tribesman was making stone blades in a forest clearing that was covered with dried leaves. He noticed a spark produced by his striking a rock which flew onto the dried leaves, igniting them. The leaves began to smoke. This revelation led the man to experiment with the stone and leaves and soon discovered he was able to kindle a flame. From that time on they had no need to transport embers from their hearths to other areas for food or warmth.

Since the tribe was always on the move, looking for places to find game and berries, they either lived in the open fields or constructed temporary crude shelters made of mud and branches. Their hearths were usually in the middle of the camp and the fire was kept alive through the night to keep predators at bay. They were very conscious of the possibility of starting a brush or forest fire from the embers left in their hearths, so they always scraped up the embers and threw them into whatever water source was available.

Asamake thought how much better things were for his group compared with his early ancestors and for some of the other clans who did not have all the social rules that were given to his clan. He was glad that his group had not lost touch with the Creator as others had done. When the clan needed guidance, the elders asked the Creator for help, and it seemed a solution always presented itself. But this was a bigger undertaking than anything they had done before, and their survival depended on it. If they did not move to wetter and greener regions soon, they would not survive as a group. In the past they moved within a few miles of familiar territory, waiting for a depleted area to recover before returning to it. Now they had no idea how far the journey would take them and their next destination. Asamake breathed a silent prayer for Mungu to guide him.

And as expected, the journey was long and arduous. They had taken a long route, but it provided more forested land along the way. In what is known as Uganda today they found the beginning of a great river, the Nile. For many months they followed it, setting up shelters in areas that provided food sources and moving on when the resources depleted. They followed the river through Ethiopia and months later arrived at its estuary where it emptied into a large sea in the Sinai desert. Their journey took them through a sandy desert, the likes of which they had not seen before, and if it was not for the river they would have turned around and returned the way they had come.

Along the way, which involved many months, the tribe continued to create advances in technology. Multiple blades were fashioned with variable edges. Human remains have been found in Blombos Cave in South Africa close to the Klasies River, in Omo, Ethiopia, and around the region of the Sinai Peninsula. What they found is that Asamake's people looked similar to current-day humans. They were the first anatomically modern humans (AMH). In terms of physical characteristics, they had a flat face, curved skull, and finely sculpted facial features of modern man. They had a hearing apparatus, and thanks to Elohim's changes to their larynx and pharyngeal muscles, they could talk to each other. In fact, a working language evolved over time, allowing them to communicate. Though there were no major physical changes in them since they left the Klasies area, they were a different people.

What changed was their ability to solve problems and innovate. This was because they now had a language to communicate and could combine the ideas and discoveries of many individuals to tackle a problem. They were by nature a confident species and invited all members of the group to contribute ideas and share in discoveries. They advanced rapidly into a complex, intelligent, and communal society that made the most of whatever resources they had. They invented the bow and arrow and honed the art of making sharp points for use in tools and weapons. They made throwing spears to aid in wounding large animals before killing them. For a while they remained connected to Elohim, who they prayed to as Mungu, the Creator and Protector. Many generations later they reached the fertile plains of the Nile in Southern Egypt and there they formed permanent settlements and farmed grains and vegetables, and continued their hunting activities.

Some of the descendants of Asamake, ever driven by man's urge to look for something new, wandered westward, populating the shore of the Mediterranean Sea as far as Gibraltar and settled there. They may have managed to cross the Strait of Gibraltar. An unusual culture called Bohunician was found in Spain and other parts of Europe which had no similarity to Neanderthal cultures. Unfortunately, despite the numerous tools found in Western Europe, there are very few human fossils. The most recent find are teeth which are different from the cultures produced by the Neanderthals in caves in Spain. The area was also occupied by Neanderthals before humans arrived. There is not a lot of evidence that they clashed or developed a close relationship. A dearth of fossils existed in the region. Most of the answers were found archeologically. Another group proceeded eastward along the Mediterranean coastline and through Anatolia, which today is known as Turkey. They continued westward and finally reached the area called the Levant. Some settled there, while a small group went south along the Euphrates River and reached the Persian Gulf.

A third group worked its way eastward through Asia eventually settling in Afghanistan. By sixty thousand years ago, most of the Middle East and Asia were settled by Asamake's descendants. Despite the advances in tools, they were still hunters and gatherers, with small attempts at farming.

Chapter Twenty-Seven

Karstagians:

Remnants of the Population That Lived Around Eden

10 kya

The group that settled in Afghanistan continued to remain attached to Elohim. He was puzzled by the fact that the various groups of humans still operated as hunters and gatherers. They had no permanent homes and built temporary shelters and stored their belongings in temporary pits in the ground. He decided to see if He could cause some momentum toward permanent towns and cities by getting one group to build a community. He chose Asamake's group in Afghanistan as they were still communicating with Him and asking for help. He talked to the current chieftain who was a young man and described the many advantages that came with having a permanent home. They were only minimally familiar with farming or irrigation but had some idea of owning animals. An intelligent young man, he was amazed that Elohim spoke personally to him. This was a reaction Elohim had never had with the other apes and australopiths with whom He spoke in the past. He would observe how the relationship developed. He told the young man that He would find a good location for them to start a farming homestead and would introduce them to some people who were experts at farming. They agreed to meet with the experts.

Elohim then recruited three of the Nephilim from His world and asked them if they would help. The meeting consisted of two men and one woman who were tall and stately. The humans appeared a little afraid of them because of their size. Together they talked about the project and the Nephilim put them at ease and soon the humans were asking questions about suitable crops, what tools they would need, and where to find seed. It seemed like a workable project, so Elohim left the Nephilim with the humans for a couple of days so they could search for a good location and find materials to build habitations. When Elohim returned two days later, they were totally supportive of the plan and wanted to start farming right away. He gave them His blessing and left, knowing that they needed to spend more time with the newcomers and get to know them.

When Elohim returned a week later, He found them working at a site they liked. It was close to a small river which flowed slowly and where silt developed. There was plenty of grass for feeding animals. They were eager to plant and harvest crops that were edible. They asked Elohim about the availability of seeds to plant for various crops as they had done in their abandoned village. The wooded area that backed up to the river provided protection for the clan. They sat down and planned on how to divide their people and organize them into two groups. One group would farm and the other would try to obtain meat in some form. Elohim told him they had done well and that things were now in their hands. If they needed help or start up seeds, they could contact Him. Elohim was optimistic that this new lifestyle was going to work and was impressed with the creature that He had created. He decided to wait on warning them about Satan.

A year later, Elohim dropped in on them and was impressed at what they had achieved. Shelters made of wood had been constructed for each couple, while two large units housed the rest, one for the women and the other for the men. The Nephilim guided them on what materials to use and devised simple ways of propping up walls and creating roofs. Looking at some of their artwork it looked like they already knew how to work with metals. They found ore in the vicinity and all they needed to do was set up a forge and then transition to bone or metal and wood tools. They asked about needing to create items like metal nails. Elohim asked His angels if they were happy with the responsibility placed on them. They loved it. He told them they were on their own and that they knew how to contact Him if needed.

A hundred years passed. Elohim had been watching from a distance but thought it was time to visit the clan in person, as He had received no calls for help in that span of time. He was impressed with what they had accomplished. They were a large clan now and had expanded to two more sites and had increased both their residential facilities and their people. Children of all ages were running around playing. He found Nephilim down by the river, checking the irrigation system which moved water from the river into holding areas on land using a pump. They no longer carried buckets back and forth to the community.

Their crops seemed to be flourishing. Chickens were running around, and they had found some goats that provided milk. They crowded around Him thanking Him for setting them up. He asked if they had neighbors yet. Apparently, they did, as there were three other homesteads like theirs. They grew different products and bartered their produce. They decided to name the group, and called themselves the Karstagians and their settlement was known as Karstag. Elohim asked if the hunting was productive, and they burst into laughter. There was no need to hunt. Between small animals that they snared, the milk products, and the chickens, they had plenty of the animal products needed to survive.

Elohim was thrilled with what they had achieved. He was worried that competing groups might start fights over ownership of property. They asked Him to stay and have a meal with them and He accepted. This was exactly what He had hoped for in His relationship with man. He thought again about warning them against Satan, but perhaps it would be better if the Nephilim did it instead, as the Karstagians trusted them. He hoped this little settlement would be duplicated all over the world as word spread to outlying areas. As time passed, the quality of life associated with a permanent residence would curb their restless desire for something better. The poor quality of their former life had prompted their endless wandering. The other factor that contributed to the wandering was the search for food. With their new lifestyle they controlled the quantity and quality of their food. Gone were the periods of hunger when all the herds migrated, and the forests turned into savannas. They had grown to love their village and were becoming more adventurous with the types of livestock they maintained and the food crops they wished to produce. They even tamed two wild dogs.

The best feature about their new lifestyle was the security in its permanence. Cold spells would occur as they had in the past, but they were prepared. The people and villages that surrounded them had already organized storage of surplus crops in underground pits in the permafrost. They planned to share the extra and, if necessary, ration food if another cold spell occurred. In preparation for the winter, the women made warm clothing of soft cured leather. He spoke to the Nephilim and asked them to warn the people about Satan and his tricks. They agreed. He left the settlement feeling a little sad. He needed to pay a visit to His angels.

Chapter Twenty-Eight

Four Thousand Years Ago:

The Cradle of Civilization

Elohim had allowed Beelzebub to maintain its power to control inanimate matter. It decided to use this skill to make life difficult for the Karstagians. It manipulated the massive ice sheets at the South Pole, causing the location of the northern end of the earth's axis to gradually increase its tilt from twenty-three degrees to twenty-four degrees northwest. The North Pole was now tilted one degree more toward the west and, as a result, the northern hemisphere had less time in the sun which made for shorter days. The steppes of Siberia became colder with longer winters as well. That winter they developed a thicker than usual layer of permafrost. Agriculture in this environment became impossible. They had to revert to hunting and foraging for food. In desperation the Karstagians asked Elohim for help through the Nephilim who still lived with them. There was no option but to move to warmer territory.

The Nephilim organized them for the move. They collected and stored smoked meat, grains, nuts, and fruit, and the women sewed together warm clothing and boots made of animal skins to keep them warm during the long trek. The men made extra tools and weapons using the metals available to them and packed raw materials to make future tools. They decided to travel in two groups. One group would move westward and find the Caspian Sea that the Nephilim told them existed to the west. The other group would travel eastward to discover what existed in that direction. Beelzebub sought to disrupt this organized move, so it caused volcanic activity to erupt in their area which caused fires and earthquakes. Many of the Karstagians were terrified by these natural events and believed that the demon gods were punishing them for trying to move. They refused to leave. Finally, with the encouragement of the Nephilim, most of them organized a trek westward and a smaller number started their trip eastward. A few remained in the old habitations hoping that things would improve climatically.

Though they were not that far from the Caspian Sea, the journey to the west would take many months. Encumbered with all their belongings, their children, and the older members, they could only move a few miles each day. The worst part of the journey was the descent from the mountains to the steppes below. Once they passed that area, their travel consisted of short days. Often, if they found a habitable spot, they would build temporary shelters and stay for a few days. Not knowing exactly where the great lake lay but using the stars and the position of the sun, they slowly moved westward, often guided by Elohim who remained connected to them. They frequently asked for and thanked Him for His help. Around 7000 BCE, the group heading west finally sited the waters of the great lake, the Caspian Sea, as it is known today. They looked for the best place to form their settlement and eventually built their village on the northwest shores of the lake. In a few years they were well on their way to re-establishing the civilization that they had known in Karstag.

Some of the group traveling to the east decided when they were halfway there that they would like to head in a southeast trajectory. They were tired of the bleakness of the Russian steppes. The remainder chose to continue to the east. This group consisted of young adults and a few older veterans who loved travel. They said goodbye to each other and continued their separate ways. The southward traveling group took a southeast approach. They traveled many days without seeing any settlements but were glad that the climate seemed to get warmer as they moved southward. They knew the ocean was to the east, so they headed in that direction. Soon they arrived at a large river. The area was sparsely populated, and the residents were friendly. With a river present, all they would have to do was obtain animals and seeds and they could farm again, after the winter ended. Elohim told them He would take care of the weather by correcting the slant of the earth back to its usual angle.

Establishing fields and having some animals should be just what they were looking for. They set up camp and tried to communicate with the local population and had a pleasant surprise. When they met new populations on previous travels, they found that most people did not have a language, so they communicated by sign language. The people where they planned to farm spoke a language, but it was different from their own. So, they reverted to drawing pictures in the sand and soon understood what they needed, as the people were also farmers.

By signals they asked the people if they could sleep at their location. They were exhausted. The people signaled to ask if they would like to eat, and they replied an enthusiastic *yes* through sign language. They set up camp as tidily as they could looking for a soft area to lay their cured skins. Looking around, they discerned that the place would be quite different come spring. They hoped they would be allowed to stay. They would offer the people their technology and co-exist with them, if they were willing. The general mood as they waited for their meal was cheerful. They spent a few minutes talking to Elohim and thanking Him. He told them they had made a good choice and would be happy there. He went ahead and returned the earth's axis to its usual angle. Warming did not begin right away since they were in the beginning of their usual winter. The shift would point the earth toward the sun and shorten the winter but would not eliminate it. The next day they discovered that they had reached the far east, and a river called the Yellow River. The changes in Homo sapiens species living in different environments had already started superficial changes in features. They were intrigued to see that the local people had different eyes with eyelids that slanted upwards. They liked the look.

The first year in their new location, they endured a severe winter and were unable to farm, but had some goats and sheep which were kept in shelters. They stayed inside and when the winter ended, they began farming the way the Nephilim had instructed them. They flourished and soon, new cities developed. They became part of the Yellow River civilization.

Meanwhile, the group traveling eastward found that the cold Siberian steppes extended for countless miles. Most of the passes through the mountains were closed, but they finally descended from the mountainous area around their village to the plains below. Once on the steppe they headed in an easterly direction. They communicated with Elohim and asked for advice. He told them to continue eastward, and it would eventually lead them to the ocean. From there they could cross over into another land. They might have to camp where they crossed if there was a lot of ice on the other side. Eventually it would melt and there would be fertile lands with a warm climate to the south. This would allow them to take their time moving through the steppes. He would keep in touch with directions. They were heartened by His words and proceeded in a northwest direction. They survived as best they could and eventually came upon the tributaries of a mighty river. They found a narrow crossing and continued over several more crossings across branches of the main river that formed a network. They stopped and rested at one of them. Once past the rivers they went northward as recommended by Elohim. Eventually they came upon a seacoast and followed it eastward until they arrived at an area around the land bridge between Asia and North America called Beringia. This was an area that existed for several hundred miles on either side of the current Behring Strait. The climate was milder since the area received little snow in the winter months and the ground was not icebound. Here they found other populations of Homo sapiens who had migrated to the same area at the beginning of the winter.

They were unable to migrate south because of the massive ice sheets that covered North America. They made camp in Beringia and exchanged stories with the established inhabitants. Elohim spoke with them telepathically and told them that the other two groups had done well and should be close to reaching their destinations. He told them that they would like the climate in the country of their destination. He advised them to travel across the land bridge and head south to the west coast of the country that they would enter. This would be possible when the ice sheets melted. If they continued going south along the coast they would eventually come to the end of the continent and would be able to walk eastward until they arrived at the location. They should ask for advice on where to farm, and find sites where irrigation was possible. The valley which they would come upon, close to the coast, would be an optimal place for farming. He would guide them to the location.

In 6000 BCE, the ice sheets began to melt, and H. sapiens sapiens wandered southward and settled in the area that is now California. Their adventurous nature did not allow them to remain in California, but they kept looking for better places to live and new sites to see. They moved along the coast until they crossed the narrow bridge between the North and South American continents and moved into the territory that is now Mexico. They found a river that had a bend with sediment accumulation, identifying it as an optimal location for establishing irrigation systems. Soon they settled in the Tehuacán Valley which was fertile land suited to agriculture, and started farming corn, squash, and beans. They were the first people to plant corn. For many generations they lived in cave dwellings, but as the ancestral memories returned, they rejuvenated cultural ways of living and eventually produced a civilization that resembled the civilization they knew in Karstag. They became known as the Tehuacán Valley civilization. They learned, by talking to neighbors, that they could grow some types of grain and vegetables. They thought of all kinds of possibilities and were very excited about the prospects of living in this new land. There were other migrants who had also recently arrived in the area and were creating their own homestead a few miles further along the estuary of the river. The first priority was to learn each other's language. They walked over to the neighboring land and obtained information about the availability of animals and seed and the possibility of bartering. There was a small town not too far away and there was availability of start-up material. The weather was warm and somewhat humid. It was summer in the area.

Elohim was pleased that the knowledge the Karstagians obtained from the Nephilim was not lost and He watched in satisfaction as new groups of civilized man arose in the areas where they migrated. Elohim watched the progress made so far by His creation and was satisfied. The initial knowledge He had transmitted through the Nephilim was passed from generation to generation in each of the groups that migrated away from Karstag. Soon, He knew, the individual groups would settle in areas dispersed throughout all the continents and grow into civilizations that would outdo the achievements of His Karstagians. Their knowledge of irrigation techniques would enable them to form settlements wherever a source of fresh water was available. Their knowledge of the architectural and structural needs of permanent settlements would enable them to build strong cities which would outlive the current generation and those who followed. The Nephilim also encouraged them to learn how to make farm tools and learn various techniques for moving water to distant fields.

Features of Himself that He endowed to man included creativity, curiosity, and the need to love and be loved. They would result in wonderful works of art, music, and technology. The possibilities were limitless if man chose to follow the moral ethic imposed on him. And yet, Elohim saw that Beelzebub, by appealing to the rights of man to have individuality and freedom, was creating a difficult situation. Though He agreed with that statement, He saw that Beelzebub had not explored the limitations on both these rights. Humans could do as they wanted as long as their actions hurt no person or unnecessarily slew animals or treated them with cruelty. If they wanted to live without help from Elohim, they were allowed. They would be judged at the end by their overall behavior during life. They could choose to be left alone and plan to lead a good life. If they did so they would be treated the same as those who lived a good life with His loving help. The smoothest path, however, was for all to live closely connected to His love and accept His guidance and assistance in all matters, every day. Soon they would do this automatically and find themselves living in a low-stress environment. Choosing His guidance would not only give them help when misfortune came upon them, but it would also give them peace of mind and knowledge of His love for them. His promise, *"Ask and you shall receive,"* is truer than most people realize. You just have to ask for reasonable things like help to resist temptation and the ability to withstand stress.

Beelzebub was disturbed by the spread of man across the world and his tendency to live in congenial communities. Above all, it feared the close relationship man now had with Elohim. Could Beelzebub overthrow this relationship? It saw only one way. In every community there were men, who like it, sought to have power over those with whom they lived. They were often outcasts, for their nature led to aggressive confrontations that were frowned upon by the elders of the groups. Beelzebub decided it would slowly empower and help man by capitalizing on the characteristic of greed that man often demonstrated. By promising them material gains, it could get them to do things that Elohim discouraged.

Beelzebub smiled as it visualized this scenario. Soon it would twist Elohim's creation into behavior that would produce the environment Beelzebub enjoyed the most—chaos. The game between it and Elohim had begun in earnest and promised to be interesting, if not exhilarating. Elohim was not perturbed. He felt that He had produced a species that so far had proved to be caring and thoughtful of others. They were living in peace with their neighbors in all the environments that they established.

Unfortunately, or perhaps fortunately, no other human species survived on the earth. The Neanderthals were the first to become extinct. Elohim decided He would give them a second chance and moved all previously existing members of the species to Haven. He felt they would have a congenial relationship with the hominins. To make this more likely, He introduced the structures that would enable speech and processing the spoken word to all the bipedal apes. He would talk to them about it in the days that followed and waited to see just how sentient they were. He had increased their brain size when He moved them to Haven, so they probably would surprise Him. They were a temporary species on Earth, but on Haven they could live as long as their natural lifespan allowed them. The sediba women had solved the problem of how to create the first Homo individuals. He created them from scratch in the wombs of four sediba mothers who willingly volunteered to care for them until they came of age. The pregnancies began naturally. The area where they originally lived is still called the hominin corridor. As a group they lived well and were on good terms with their neighbors in South Africa and Malawi. Only three of them lived a long life. The rest faded away and no further information about the cause of their extinction is known by anyone other than Elohim, who whisked them away to a planet called Haven to live happily thereafter. It is possible that they will produce a Homo-like species there.

The Homo heidelbergensis species in Africa and Eurasia had also silently disappeared after their responsibilities to produce two new species and organize the Homo species in Europe were completed. Elohim felt that they had gradually come to think of themselves as H. sapiens and blended into the AMH population.

Homo ergaster/erectus species existed for almost a million years but became extinct 117,000 kya to 80 kya. They were devoted to Elohim and, as a result, they did not need to live in large groups. They formed small family groups and did not increase in numbers to the same extent as other Homo species. Elohim had arranged that members of both species would be transferred to Haven when they died so they could be with the sediba population. Some paleontologists believe that after the H. ergaster-to-erectus transition was completed through the *grade transitions*, each of the grades retained their final morphology and remained in the areas where they had evolved. Homo erectus species successfully produced many of his kind and they spent their lives telling new humans about Elohim and how to serve Him until they died. Erectus developed a more gracile skull and mandible but otherwise did not advance to AMH. Elohim had already transferred both erectus and ergaster families, including the ones that had died, to Haven.

Homo antecessor did not survive long in Eurasia. While H. naledi evolved and lived for a short while in South Africa, he too became extinct 335 kya to 236 kya. The fossilized remains of many members of his species were recently found in a cave laid out on natural ledges. No personal property in the form of tools or personal items were left with them. They are assumed to have been descendants of early Homo ergaster that were still living in South Africa. However, the investigation into their origins has barely just started. One of the interesting things about his anatomy was that he had a very small skull and many antique features. His arms and legs were those of an erect species, but he could also climb trees. With so many intact skeletons to work with, it is anticipated that an explanation as to when and where he evolved will be available in the near future. The priority is to find viable DNA. They have made numerous attempts to obtain viable DNA from different parts of different bodies, but all have turned out to be destroyed by aging and a hot climate. The climate in South Africa is very hot in the summer, so the chance of having intact DNA is quite low. Researchers still have a lot of material from unexplored body parts to untouched skeletons to review. A second site was found in an adjoining cave, and they are working on retrieving the fossils in that location. It is possible that temperatures deep inside the cave were lower and will allow for intact DNA in the younger fossils.

Chapter Twenty-Nine

The Denisovans

200 kya–50 kya

The only other species that was newly discovered is the Denisovans. Their initial remains were found in a cave on the edge of the Tibetan Plateau in Gansu, China. Unfortunately, there was only one fossil, a finger belonging to an adolescent female. This was studied using mtDNA and nuclear DNA. The Tibetan Plateau is very cold most of the year, so it is not surprising that the DNA would be well preserved. The nuclear DNA showed a close affinity with H. neanderthalensis and some evidence of overlap with H. sapiens genes.

The next fossil was found in a different cave by a Buddhist monk. It was half of a mandible, followed by small bones and a few teeth. Over the next few years, several teams of experts worked with Chinese scientists to try to draw a picture of what the species looked like and where else it had traveled. The possibility of human DNA in their genome was of great interest. It was examined and found not to be like any of the mandibles that Homo species have. There was no chin, which is no surprise since only H. sapiens sapiens have a chin. The mandible was donated to Lanzhou University and sat on a shelf until 2010. By that time new DNA techniques had evolved, and scientists were able to complete the DNA processing of tissues. The new processes were less harsh in the handling of the partially damaged and very old DNA.

A total of five sets of fossils were found in China, except for the mandible, which was found in a different cave system, the Denisovan Cave, located in Russia. The floor of the cave was dug to different levels and the contents examined. Objects eligible for age dating were processed with a variety of radiometric techniques at each level of stratigraphic study. Other than the mandible and the finger, all fossils were small and were in the form of fragmented bones or whole teeth. No skulls or other intact skeletal fossils were found or have been found anywhere else. All information about the creatures comes from the study of their DNA. Due to the lack of evidence, there is no speculation regarding what the creatures looked like. Recently, with a new dating technique it became possible to make predictions of what the species might have looked like. A model of a Denisovan head and neck are available to view on the Smithsonian Institute website. It provides the possible appearance of the Denisovan species and a visualization of other Homo species. The latter are likely to be representative of the original, since skeletons and skulls were available and tissues could be molded using a clay-like substance to determine what they might have looked like. This is a solitary study, and researchers await confirmation from other studies. Pictures are presented on the Smithsonian website, but scientists are not able to confirm these results other than waiting for further work using this technique.

Other fossils found include the distal end of the fifth finger of a thirteen-and-a-half-year-old girl found in 2008. This was tested for mtDNA and nuclear DNA as outlined above. In 2010, permanent upper second or third molars were discovered, and they came from an adult male. Fortunately, the arrival of this species, coinciding with the appearance of Neanderthals, occurred recently enough that scientists were able to study their DNA and obtain more information about them. Homo heidelbergensis was the ancestor of both Neanderthals and Denisovans. To refresh his history: H heidelbergensis diverged seven hundred thousand years ago from an unknown ancestor which emerged eight hundred thousand years ago. Soon after, H. sapiens split from the same lineage. Researchers do not know this ancestor, but judging by the information gleaned from both the Denisovan genome and the mitochondrial antigen found in the female's finger, it is thought to be intellectually advanced for the time that she lived. Denisovans, Neanderthals, and humans occupied the Denisovan cave sequentially. This is known because fossils of each species occupy separate layers with no admixture of another species or its cultural objects. So, none of the species had contact with one another.

To summarize, the genetic study of fossils found in the floor of the Denisova Cave were tested genetically by genome, protein, and mtDNA analysis. The search for fossils was done stratigraphically and four layers were found and examined. Each layer produced many small fossil fragments and teeth. Also present, were small items that identified which species may have lived in the cave. The deepest layer and therefore the earliest occupation was by the Denisovans. Tests for their age revealed that they were there 160 kya to 20 kya. Three more layers of body parts were found and identified from forty thousand, sixteen thousand, and eight thousand years ago. These discoveries all look like human occupations with tools and other artifacts belonging to Homo species including Homo sapiens, Homo neanderthalensis, and Denisovans. It does not appear that at that point in the timeline any other species cohabited with the Denisovans at this location. Each species was in the cave independently at consecutive times and had no contact with each other. There was, however, evidence of Denisovan DNA in the H. sapiens genome. This must have happened before they visited the cave.

The origination of Neanderthals from H. heidelbergensis occurred about seven hundred thousand years ago. It is not clear where this occurred, but it had to be somewhere in Europe because H. neanderthalensis did not reach Africa. No fossils have been found in Africa and it is assumed to not have spent time there. The recent findings at Atapuerca, Spain, show fossils of Neanderthals, H. heidelbergensis, and H. antecessor in two different dig sites. No evidence of H. sapiens is present with them. The timing for the site is three hundred thousand years ago and there is no sign of the intelligent Homo species that gave rise to heidelbergensis and Homo sapiens seven hundred thousand years earlier. It is assumed to be extinct.

Elohim drafted heidelbergensis to perform oversight of AMH two hundred thousand years ago. This means that heidelbergensis was in Africa by two hundred thousand years ago. Some scientists think there were two versions of heidelbergensis, an African and a European. This was because Elohim enjoyed His first contact with the species and felt he was perfect for talking to people and telling them about Him. He appointed a second contingent of the heidelbergensis clan to live in South Africa at the Cape with a group pf archaic H. sapiens that already lived there. The heidelbergensis He chose settled down at the tip of South Africa and got to know the group of archaic H. sapiens that lived there. He had also proved to be very capable, so they were not two separate groups of heidelbergensis, but two individuals in the same group. One met and socially interacted between 200 kya to 160 kya with Homo groups and Denisovans in Eurasia. The other stayed in Africa and met the different groups of AMH as they went back and forth from South Africa to North Africa to escape seasonal extreme temperatures.

Paleontologists believe that the Denisovans left the Denisova Cave and moved to the Oceana Islands fifty thousand years ago. They were in Papua New Guinea, Australia, and possibly other small islands in the Pacific. They may have become extinct twenty thousand years ago or became absorbed into the tribal populations with whom they lived. It appears that they also spent some time in the Philippines before moving to the Oceana Islands. A recent DNA study on native tribes living in the Philippines showed that the Ayta Magbukon group that currently lives there has a higher percentage (up to 40 percent more) of Denisovan DNA than the percentage seen in the Papuans. This suggests that the Denisovans spent some time in the Philippines before moving to the Oceana Islands.

Further work on the findings at Atapuerca may still yield some clues as to how the Homo species made contact with the others.

Chapter Thirty

The Indus Valley Civilization:

The Great Flood

6 kya

The Karstagians lived on the shores of the Caspian Sea until about 6000 BCE. As they became prosperous, they began to trade with the people of adjacent territories. Their traders traveled into the Levant and exchanged agricultural products and bronze artifacts for salt and obsidian. They also traveled eastward and made their way to northern India. On the banks of the Indus River, they found people who farmed the land and kept animals. Their technology was not as advanced as that of the Karstagians, but they settled in the area and taught them their crafts. In a few thousand years, they shared a language and became known as the Harappans. They built cities with modern amenities. Mohenjo-Daro, the area in which they lived, and its ruins still exist today. They are called the Indus Valley civilization.

Meanwhile, the people living around the Caspian Sea saw an increasing flow from the rivers from the north resulting from the melting of Scandinavian ice sheets. Simultaneously, the water levels in the Atlantic Ocean extended to the Mediterranean Sea, which filled up and began to spill over its shoreline. The combination of the two natural occurrences affected the area around the Caspian and Black Seas resulting in massive flooding. The Mediterranean Sea topped a spur of rock that contained it and spilled over as a cascading waterfall which fell at a high speed into the Black Sea in the region of the Bosporus Straits. The straits connect the Black Sea to the Sea of Marmara. The flow continued for months. This transmitted story probably has a measure of exaggeration in it.

Elohim, foreseeing the flood, warned some of the tribal elders who lived in the area to move south with their belongings and animals and to build boats that allowed them to cross the flooded rivers and lakes. Sometime around 6000 BCE, the descendants of the Karstagians started the trek southward moving away from the flooding areas to warmer lands. The people were grateful for Elohim's warning which gave them a chance to prepare.

Their journey southward brought them to the area that was Mesopotamia, now known as Iraq. There they found two rivers, the Tigris and the Euphrates, which were exactly what they needed to continue their farming activities. In this new land, they started farming and rearing goats and cows for milk. The silty soil of the estuaries of the two rivers was rich and fertile. They thanked Elohim for directing them to this destination and vowed to work hard, making it productive for their people and to share with others.

The story was told by verbal repetition to young children and relatives and eventually became a part of their mythology. In about 3500 BCE, they invented cuneiform, an ancient form of writing. They were the first people to invent a true form of writing. It was not pictographic, and it was not a language. One of the tasks assigned to the scribes was to write the legend of Gilgamesh. A few thousand years later it resulted in the great epic poem describing how Gilgamesh, a king who received a message from the creator, warned his people about the coming flood. It described the measures needed to be taken to save lives. The story of the warning by their god who saved their lives was preserved in these ancient people and led to stories of the great flood that became a well-known legend to all the neighboring tribes as well. The writing script was easy to use and was modified and adopted by many cultures in Mesopotamia that came after the Sumerians.

To date no one knows from where the Sumerians migrated. There are two main theories based on their writings, mythology, and the similarity of their written language to present-day languages. Their writings speak of a homeland called Dilmun and some anthropologists feel that this may have been one of the islands in the current gulf of Arabia. The people who were settled in the area and in adjoining Iran are thought to have traveled by sea in boats from the west coast of India. They are thought to have been of the Dravidian stock that inhabited the area comprising India and Pakistan at that time. Another theory is that they arrived from the region around the Caspian Sea. The possibility that they ultimately were the descendants of archaic H. sapiens that lived in the area of Eden in South Africa is not prevalent. Nor is their connection to the Karstagians who lived in Afghanistan.

The original people were called Subarians in various written accounts and were probably the precursors of the people who are known as the Sumerians. Although neither theory has been proven, the language similarities indicate common roots for these migrants and for the people currently living in South India, Iran, Finland, Siberia, Iceland, and parts of South America. There are fundamental similarities in all these languages that are completely different from any of the major language groups that exist in Europe today.

Alfred Hamori, a Hungarian scholar of Sumerian mythology, postulates that the original Sumerians were an ancient and advanced civilization that was forced to migrate because of the sudden onset of freezing conditions in Siberia. In his article, "The Origin of Sumerians and the Great Flood," he suggests that the polar shift in 12,000 BCE caused a major climatic change in the steppes of Siberia which progressed from a moderate temperate to an arctic climate. To escape from the cold, the people living in the area migrated in different directions. Some crossed the Behring Straits into North America. Others moved west into Central Europe and some further south to the area around the Mediterranean. Massive North American ice sheets developed in the one thousand years of glaciation that followed the eruption of Mount Toba seventy-four thousand years ago. The continuation of the ice age, which peaked at 12,000 BCE, added to the initial mass of ice at the North Pole which, at the time, was supposedly located in the Hudson Bay area. Over the next two thousand years (ending in 10,000 BCE), the massive ice sheets in the area slowly melted and, as a result, sea levels rose along the coastline of Europe which gradually became flooded. By 10,000 BCE, the British Isles were separated from mainland Europe by the Atlantic Ocean. A similar flooding of the Mediterranean coastline supposedly wiped out cultures living on its shores, resulting in the Egyptian myth about the buried city of Atlantis.

Simultaneously, the level of the Black Sea, which was then a freshwater lake, decreased so that it was below sea level. Around 5500 BCE, the sea level of the Mediterranean rose high enough to breach the heights overlooking the basin of the Black Sea. The end result was a sudden flow of the Mediterranean from a great height into the Black Sea.

The overflow of the Mediterranean into the Black Sea suddenly and forcefully is supposed to have occurred around 6500 BCE. The science quoted is from a study done by Pitman and Ryan in 1998. They wrote about it in their book, *Noah's Flood: The New Scientific Discoveries About the Event That Changed History.*

In this book they describe their findings from a geologic expedition conducted by the U.S. and Russia as a joint archeological expedition. Their hypothesis was that this kind of sudden overflow of the Mediterranean Sea could not occur in the baseline system of a large sea with limited capacity to expand with good connections to multiple bodies of water (the Black Sea and the Caspian Sea) acting as buffers. It also mitigated things because both the Caspian and the Black Seas were below sea level. They did an extensive study of the floor of the Black Sea where the overflow took place, examining and taking samples from several biospheres, extending into the shelf below sea level. Obtained shelf cores showed that penetration below shelf level had occurred which indicated that the water fell with some force. Their conclusion was there was a possibility that the Mediterranean spilled over into the Black Sea. However, the extent and force of the overflow may have been exaggerated. Similar studies had opinions on this issue, but none gave conclusive evidence. The issue therefore is undecided. No evidence of Atlantis-type ruins of habitations was found.

While living together on the shores of the Caspian Sea, considerable cultural exchange had occurred between the two groups that lived there, resulting in similarities in language and mythology. Hamori also postulates that some of the people who were displaced by the flood moved northeastward and settled in the areas which are today Finland, Transylvania, and Hungary. This again is inferred by similarities in language and mythology. The myths in all three new habitations describe a homeland which was warm and pleasant with seven months of summer. They describe a time when the climate changed and became extremely cold with ten months of winter, and this was blamed in on evil gods. And they describe a high flat mountain in the area where they lived. Possibly this caused awakening of their memories of Table Mountain in South Africa.

Based on their mythology, Hamori believes that the original homeland must have been in Siberia and that the climatic changes described were due to the polar shift. He explains the later primitive culture of the Siberian population as being due to their isolation in extremely inhospitable environments. Eventually memories of the previous civilization were lost. The people who managed to move to warmer climates, as they existed around the Caspian Lake, carried their culture with them and it flourished in the new environment. Many historians look at this time interval as a time when hunter-gatherers were experimenting with small in-place farming and a flood would surely have created havoc with such ventures. This should have resulted in some form of documentation later on. It has not apparently occurred.

One explanation is that the icing of Eastern Europe saw large migrations from that area to Western Europe. Their farming ventures were destroyed, and many moved to areas that had potential for farming, such as lake shores and river estuaries. It may be that the immigration of East Asian residents looking for land to farm had been more important or interesting. In so far as we know, writing did not exist six thousand years ago and apparently the difficult journey each group had experienced did not get recorded when it emerged thirty-five hundred years ago. They may have had recent material to write about. Also, people with bad experiences do not always like to perpetuate the memory of them.

This was a time when farming cattle as well as growing crops dramatically increased. The domestication of cows was first achieved in the Far East. From there the cows were imported to Eastern Europe and introduced to other areas that were suitable throughout the rest of Europe and the Middle East. There were very few human populations that existed at that time. So, the small groups could still find areas for farming and were not forced to return to foraging and hunting.

This narrative of the Sumerian's origin has some backing from archaeological finds in Transylvania, Romania. Clay tablets showing a Sumerian type of written language were discovered in 1961 in the region. Carbon dating of these tablets shows them to be from 4500 BCE almost a thousand years earlier than those found in Lower Mesopotamia. There is also some anthropological basis in comparative skull anatomy for the common origin of the Fin-Ugor peoples, the Mongolian peoples, and possibly even the Dravidian people who lived in India. They may have lived in Mohenjo-Daro when the Sumerians traded with the Indus Valley population. This occurred when they settled in the northwest corner of the Caspian around 7000 BCE, long before the overflow of the Mediterranean emptied itself into the Caspian Lake. Hamori discusses the similarities in the root words of each group that blended into their eventual language. In addition, their mythologies and religious figures show considerable overlap and there is frequent reference of the existence of demigod-like creatures with whom they associated and who protected them and taught them how to make a good life. There is overlap in their creation mythology which paints humans as the creation of a deity who created them to serve the gods.

The origin of the culture in Siberia is still mostly conjecture, supported by some fossil finds. It did not develop into the first large civilized empire, but they were the first step toward it. Elohim had been tracking them and was happy to find a population that had existed around Eden. When Adam and Eve left Eden and settled close to its proximity, the people who lived around there, most of them archaic H. sapiens, mated and stayed with them. Elohim felt that they were at least in small part the direct descendants of Adam and Eve. They remained in the area after many neighbors had already left and followed Elohim's guidance in settling down in Afghanistan.

In a short period of time, after they moved from the Caspian shores and settled along the banks of the Euphrates and the Tigris Rivers, a civilization appeared which was in part city-oriented and partially involved in agriculture. This is called a city-state civilization. Soon the riverbanks were used for crop growing and animal husbandry. The administrative branch or government of this entity was evolving just as much as the social life of the people. Those with a knack for organizing moved into governing. They built cities with walls and gates where most people, including the farmers, lived. This was for safety. Already little power groups were forming in Western Europe which in a century or so would become warring groups.

The Sumerians were more interested in protecting themselves and establishing trading relationships. The farming community in Central Europe who were the ancestors of the current culture in the Iraq area were forgotten. No information regarding their existence in the Afghanistan area is available. Any writings by them came well after thirty-five hundred years ago and it is likely that advancing generations had lost most of their early history. It could be that the trip from Afghanistan to the Caspian Sea was so grueling that they suppressed those memories and did not want to record them as family events. They were simple farmers isolated in Afghanistan and, until writing was created, there would be no record of their history. Even the history of Eden was forgotten by the descendants of Asamaki who had so bravely set forth from the Olduvai Gorge to find a new life. In later writings there was no mention of those events either. Nor did they ever record the grueling trip from Afghanistan in the middle of a freezing winter. However, archeological exploration of the Siberian wilderness is in its infancy and findings in the future may shed light on this group.

Again, with His omniscient vision Elohim could see that there would be many difficult times and trials for man. Some of them via Satan, some self-driven, and others from happenstance. Their contact with Him would be tenuous at times but the moral code would prevail and keep mankind from completely aligning with Satan's agenda. He looked forward to the challenge. Satan had no restrictions on his attempts to bend humans to his will. He was not allowed to take the life of a human, but he could make it painful. Elohim thought of the soul, a small part of His own essence that every human received at the time of birth. Through it He could be a voice stirring up the conscience of a tempted human and inducing him to take the right path and ask for help.

When this happened, He would honor the covenant He had made with Adam and Eve and help by providing ideas on the next steps to keep the relationship intact. Elohim would give them advice regarding what to do if Satan was unfairly targeting them, when man asked for help. He would then reprimand Satan or make it possible for them to have a positive attitude and endure a temporary discomfort or unhappiness. He would miss the many informal and personal conversations He had with apes and early humans. Now that man was established, Elohim would remain in the background. He knew there would be no shortage of individuals who would connect, love, and respect. There would be men and women who would not need miracles to believe in His existence and Elohim would love them for it. He was still in touch with the apes and hominins on Haven and frequently visited with them and will continue to do so.

Acknowledgments

I owe the completion of my book to my publisher and editor, Julie Castro. She allowed me to go at my own pace and put up with numerous edits by me to make it be exactly what Elohim wanted. I thank her for her patience and her experienced advice. She also calmy rescued me out some computer glitches which I got myself into. Thank you, Julie, for your tolerance and readiness to help.

ENDNOTES AND SOURCES

Reference Books

BN1 The Holy Bible, The Confraternity of Christian Doctrine, *The New American Bible 1970, Revised Edition* (Oxford University Press, 2011).

BN2 Walter Isaacson, *Einstein: His Life and Universe* (Simon & Schuster, 2008).

BN3 Monroe W. Strickberger, *Evolution, Third Edition* (Jones & Bartlett Learning, 2005). *Begins with an interesting discussion on the conflict between religion and science on this subject. See chapter four: "The Darwinian Evolution and Religion."*

BN4 DK Publishing (Dorling Kindersley), *Prehistoric Life: The Definitive Visual History of Life on Earth* (DK Publishing, 2012). *Spectacular projected illustrations of all types of species with brief notes from prominent paleontologists. Brings prehistoric times closer to reality. Also includes major advances in evolution and visuals of the solar system. Worth looking for. Wonderful experience via images. Book may be available in your local library.*

BN5 Lisa Bartee and Catherine Creech, *Principles of Biology* (Open Oregon Educational Resources, 2017). https://openoregon.pressbooks.pub/mhccmajorsbio/.

BN6 Roger Levin and Robert Foley, *Principles of Human Evolution, Second Edition* (Blackwell Publishing, 2004).

BN7 Suzanne Wakim and Mandeep Grewal, "Libre Texts Biology 23.3: Embryonic Stage," LibreTexts.org, accessed June 28, 2025. https://bio.libretexts.org/Bookshelves/Human_Biology/Human_Biology_(Wakim_and_Grewal)/23%3A_Human_Growth_and_Development/23.3%3A_Embryonic_Stage.

Chapter One – The Beginning

BN2 Walter Isaacons's *Einstein: His Life and Universe.*

BN2 Albert Einstein's childhood and the formation of his awareness of a "Divine Spirit" in nature who does not interact with us (see "The Light-Beam Rider" in chapter 1, pp. 1–7).

BN2 Childhood (chapter 2, pp. 8–32).

BN2 "The Miracle Year"; quanta and molecules (chapter 5, pp. 75–94).

BN2 Newton's laws of motion for "absolute space and time" changed by Einstein; "The Miracle Year" (chapter 5, pp. 90–106).

BN2 Special relativity (chapter 6, pp. 107–139).

BN2 General relativity (chapter 9, pp. 189–224).

BN2 Einstein's God (chapter 17, pp. 384–393).

BN2 Einstein's mind and Einstein's brain (epilogue, pp. 544–551). The epilogue reveals Einstein's opinion on many controversial issues in the form of quotes from other people who knew him well.

Online References

*Note: The links below were active at the time of writing. However, if discontinued, the content can be traced by searching the titles and headings listed.

1. "The Big Bang," Wikipedia, the Free Encyclopedia. https://en.wikipedia.org/wiki/Big_Bang. Includes timeline, inflation and baryogenesis, and cooling. Discusses related issues: baryon asymmetry, dark energy, dark matter, and globular cluster age in physics. Concludes with horizon problems.

2. "List of Particles," Wikipedia, the Free Encyclopedia. https://en.wikipedia.org/wiki/List_of_particles.

3. "Planck Epoch," Wikipedia, the Free Encyclopedia. https://simple.wikipedia.org/wiki/Planck_epoch.

4. "Newton's Laws of Motion," Britannica, 2025. https://www.britannica.com/science/Newtons-laws-of-motion.

5. "Timeline in an Inflationary Big Bang Cosmology," Britannica, 2025. https://www.britannica.com/science/Newtons-laws-of-motion.

6. "History of Special Relativity," Wikipedia, the Free Encyclopedia. https://en.wikipedia.org/wiki/History_of_special_relativity.

7. "Electromagnetism," Wikipedia, the Free Encyclopedia. https://en.wikipedia.org/wiki/Electromagnetism.

8. "Solar Eclipse of May 29, 1919," Wikipedia, the Free Encyclopedia. https://en.wikipedia.org/wiki/Solar_eclipse_of_May_29,_1919.

Chapter Two – Heavenly Unrest

Wisdom 2:23–24 (NABRE): "For God formed us to be imperishable; the image of his own nature he made us. But by the envy of the devil, death entered the world, and they who are allied with him experience it.

Satan, Elohim's chief archangel, plots a revolt.

Chapter Three – The Banishment of Lucifer

Isaiah 14:12–13 (KJV): "How art thou fallen from heaven, O Lucifer, son of the morning! how art thou cut down to the ground, which didst weaken the nations! For thou hast said in thine heart, I will ascend into heaven, I will exalt my throne above the stars of God: I will sit also upon the mount of the congregation, in the sides of the north."

Satan attempts to overthrow Elohim and take control of heaven, but is deprived of his prize by Elohim simply moving heaven with His loyal angels into another dimension. Satan and his followers are left to fend for themselves in the material world.

Chapter Four – The Milky Way Galaxy

BN3 Monroe W. Strickberger's *Evolution* (see chapters 5 and 6).

BN3 The origin of the universe (chapter 5, pp. 79–84).

BN3 The earth (chapter 6, pp. 85–108).

Online References

1. "A Visual Guide to Our Solar System [Infographic]," EarthHow, 2025. https://earthhow.com/solar-system-facts/.

2. "Solar System," Wikipedia, the Free Encyclopedia. https://en.wikipedia.org/wiki/Solar_System.

3. Natalie Wolchover, "Cosmologists Debate How Fast the Universe Is Expanding," Quanta Magazine, August 8, 2019. https://www.quantamagazine.org/cosmologists-debate-how-fast-the-universe-is-expanding-20190808/.

Chapter Five – The Beginning of Life: DNA and RNA

BN5 Lisa Bartee and Catherine Creech's *Principles of Biology* (free downloadable book). Book does not have chapter numbers. It has headings for each new topic. The headings are linked to page numbers in the contents at the beginning. Best searching is via the contents list which is very detailed. For access to contents, click on blank upper left-hand corner.

BN5 The living cell (pp. 267–310).

BN5 The cytoplasm (pp. 78–83).

BN5 Cells: the living units (pp. 87–91).

BN5 The genetic code and DNA replication (pp. 554–558).

BN5 Protein synthesis (pp. 558–575).

BN5 Cell growth (pp. 88–91).

Online References

1. Patrick Forterre, Jonathan Filée, and Hannu Myllykallio, "Origin of Evolution of DNA and DNA Replication Machineries," The Madame Curie Bioscience Database [Internet] (Landes Bioscience, 2000-2013). https://www.ncbi.nlm.nih.gov/books/NBK6360/.

Chapter Six – Meiosis and Sexual Reproduction; Body Plans; First Eukaryotic Cell Ancestor (FECA)

BN5 Lisa Bartee and Catherine Creech's *Principles of Biology* (free downloadable book).

BN5 Meiosis and sexual reproduction (pp. 629–638).

BN5 Overview of meiosis (pp. 634–638).

BN5 Meiosis I (pp. 639–649).

BN5 Meiosis II (pp. 650–654).

BN5 Comparison of MI and MII (pp. 654–657).

Online References

1. Brian K. Hall, "Evo-Devo: Evolutionary Developmental Mechanisms" (Department of Biology, Dalhousie University, Halifax, Nova Scotia, Canada, 2003).
https://pubmed.ncbi.nlm.nih.gov/14756324/.

2. Jaume Baguñà and Jordi Garcia-Fernàndez, "Evo-Devo: The Long and Winding Road" (Department of Genetics, Faculty of Biology, University of Barcelona, Spain, 2003).
https://pubmed.ncbi.nlm.nih.gov/14756346/.

3. Lee Altenberg, "The Evolution of Evolvability in Genetic Programming" (SFI Working Papers, Santa Fe Institute: Institute of Statistics and Decision Sciences, Duke University, 1994).
https://sfi-edu.s3.amazonaws.com/sfi-edu/production/uploads/sfi-com/dev/uploads/filer/99/fa/99fa73e5-7d46-4cd2-9e2d-4c1eed56a25f/94-02-007.pdf.

4. Jeffrey P. Tomkins and Jerry Bergman, "Developmental Gene Regulatory Networks – An Insurmountable Impediment to Evolution" (Journal of Creation, 2018), 32(2)96–102.
https://creation.com/developmental-gene-regulatory-networks.

Chapter Seven – The Cambrian Explosion of Life

*Author's note: Most of this chapter is written on the assumption that the Cambrian explosion was engineered by Elohim. The articles below give alternative unproven explanations for the event. An exception is the last article by Steven Jay Gould, which paints a large picture but admits the event is not explained scientifically up to that time. The Precambrian, featuring the Ediacaran flora, is discussed by me in the context of Elohim being the one responsible for their creation. Two other views are listed. No definite scientific explanation exists to date.

BN3 Monroe W. Strickberger's *Evolution.* This book outlines the rapid evolution of early creatures in the Cambrian Era. An important addition was the evolution of metazoans. These were creatures that are supposed to have evolved from the Ediacaran fauna. All evolved in the ocean and all were invertebrates. (See chapter 14, pp. 323–342.)

BN3 From protozoa to metazoa (chapter 14, pp. 323–334).

BN3 The Cambrian explosion (chapter 14, pp. 323–328).

BN3 Protist ancestry (chapter 14, pp. 328–329).

BN3 Hypothesis of metazoan origin (chapter 14, pp. 329–336).

BN3 Forward from multicellularity; the coelom/metamerism (chapter 14, p. 336).

BN3 Evolutionary solutions to problems of locomotion (chapter 14, pp. 338–442).

BN3 The coelom (chapter 14, pp. 336–337).

BN3 Metamerism: segmentation of the body oriented in a dorsal, ventral plane (chapter 14, pp. 336).

Chapter Eight – Ordovician Mass Extinction: Satan's Attempt to Destroy the Earth

BN3 Monroe W. Strickberger's *Evolution* (see chapter 15, pp. 345–362).

BN3 Differentiation and the evolution of development (chapter 15, pp. 345–363).

BN3 Viral and bacterial development (chapter 15, pp. 345–347).

BN3 Eukaryotic development (chapter 15, pp. 347–351).

BN3 Genetic control of embryonic space (chapter 15, pp. 351–353).

BN3 From Hox genes to zootypes (chapter 15, pp. 353–355).

BN3 Developmental interactions and constraints (chapter 15, pp. 355–357).

BN3 Heredity and developmental constraint (chapter 15, pp. 357–358).

BN3 Adaptive constraint and its modification (chapter 15, p. 358).

BN3 How genotypes are "canalized" (genetically constrained) (chapter 15, pp. 358–359).

BN3 From constraints to innovations (chapter 15, pp. 360–362).

Online References

1. Brian K. Hall, "Evo-Devo: Evolutionary Developmental Mechanisms" (Department of Biology, Dalhousie University, Halifax, Nova Scotia, Canada, 2003). https://pubmed.ncbi.nlm.nih.gov/14756324/.

2. Jaume Baguna and Jordi Garcia-Fernàndez, "Evo-Devo: The Long and Winding Road" (Department of Genetics, Faculty of Biology, University of Barcelona, Spain, 2003). https://pubmed.ncbi.nlm.nih.gov/14756346/.

3. Lee Altenberg, "The Evolution of Evolvability in Genetic Programming" (SFI Working Papers, Santa Fe Institute: Institute of Statistics and Decision Sciences, Duke University, 1994).
https://sfi-edu.s3.amazonaws.com/sfi-edu/production/uploads/sfi-com/dev/uploads/filer/99/fa/99fa73e5-7d46-4cd2-9e2d-4c1eed56a25f/94-02-007.pdf.

4. Jeffrey P. Tomkins and Jerry Bergman, "Developmental Gene Regulatory Networks—An Insurmountable Impediment to Evolution" (Journal of Creation, 2018), 32(2)96–102. https://creàtion.com/developmental-gene-regulatory-networks.

5. "Cambrian Explosion," Wikipedia, the Free Encyclopedia. https://en.wikipedia.org/wiki/Cambrian explosion.

6. "The Paleozoic Era," Fossils, Facts and Finds, 2021. https://www.fossils-facs-and-finds.com/paleozoic_era.html.

Includes the Cambrian, Ordovician, Silurian, Devonian, Carboniferous, and Permian Eras.

Chapter Nine – Devonian Period: From Chordates to Early Amniotes

BN3 Monroe W. Strickberger's *Evolution.*

BN3 Origin of vertebrates; deuterostomes and protostomes to bony fish and Sarcopterygii; two pre-amphibian stages (chapter 17, pp. 391–433).

BN3 Hypotheses on the origin of vertebrates. By the time vertebrates appeared in the fossil trail they had typical chordate features. No ancestral fossils are available. Molecular studies done today using fossils suggest that an annelid or arthropod may have been the transition form. The most recent hypothesis is that they were derived from an echinoderm/vertebrate ancestry. This is based on molecular studies and projections from analyzing the embryology of annelids and arthropods. The theory is that as Metazoa split up, they produced Echinodermata, Hemichordate, and Chordata. This is based on an RNA sequence analysis which produced triploblastic homeobox-containing genes to produce the phylum which was moved from a protist group to a Metazoan group. No fossils are available to confirm this theory (chapter 16, pp. 366–368).

BN3 Cephalochordates and urochordates. Cephalochordates are filter feeders and have several features well discussed in chapter seventeen. The conclusion from embryological studies favors the cephalochordates as precursors of vertebrates. Three features that suggest that they were the precursors of chordates were the continuation of origin of vertebrates. They suggest that Metazoa evolved in the Cambrian or Precambrian seven hundred million years to one billion years ago (the same time span in which Elohim was improving genetic features in a group of sea creatures in the Ediacaran!) (chapter 17, pp. 395–398).

BN3 Fossil jawless fish (Agnatha) were the earliest vertebrates that existed in the late Cambrian. Jawed vertebrate species appeared in the Silurian. They were found only in fresh water. They evolved into Acanthodians and Placoderms in the Devonian and rapidly became extinct (chapter 17, pp. 398–401).

BN3 Chondrichthyes and Osteichthyes were the next possibility to arise. One had a bony skeleton, and the other had a cartilaginous skeleton. A number of Osteichthyes evolved a variety of directional fins. Their positions varied and they were used for changing direction in the water. They then developed several body parts being changed to cartilage. Osteichthyes converted skull, jaws, gill cover, scales, vertebrae, and ribs to bone. Chondrichthyes did the same but to cartilage. Lineages of Osteichthyes over time developed variations in positioning of fins and some did better than the others. A variety of breathing features occurred to enable higher water content of oxygen including the evolution of "lung-like" structures. Bony fish eventually evolved. They were called Actinopterygii and Sarcopterygii (chapter 17, pp. 402–404).

Online References

1. "Silurian," Wikipedia, the Free Encyclopedia. https://en.wikipedia.org/wiki/Silurian.

Has a very good timeline of life from the Hadean to the end of the Quaternary.

2. "Lungfish," Wikipedia, the Free Encyclopedia. https://en.wikipedia.org/wiki/Lungfish.

3. "Eusthenopteron," Wikipedia, the Free Encyclopedia. https://en.wikipedia.org/wiki/Eusthenopteron.

4. "Synapsida," Wikipedia, the Free Encyclopedia. https://en.wikipedia.org/wiki/Synapsida.

5. "The Permian Period: The End of an Era," Fossils, Facts and Finds, 2021.
https://www.fossils-facts-and-finds.com/permian-period.html.

Chapter Ten – Amniotes to Placentals

BN3 Monroe W. Strickberger's *Evolution* (see chapter 18).

BN3 "From Water to Air: Amphibians, Reptiles, and Birds" (chapter 18, pp. 409–438).

BN3 Early amphibians (chapter 18, pp. 410–414).

BN3 Modern amphibians (chapter 18, pp. 409–416).

BN3 From amphibians to tetrapods to amniotes (chapter 18, pp. 416–419).

BN3 Reptilian evolution (chapter 18, pp. 419–425).

BN3 Diagram of amniotic evolutionary tree (chapter 18, p. 422).

BN3 Early archosaurs (Permian) (chapter 18, pp. 424–425).

BN3 The dinosaurs (Triassic) (chapter 18, pp. 425–427).

BN3 Endothermy versus ectothermy (chapter 18, pp. 427–429).

BN3 The late Cretaceous extinctions (chapter 18, pp. 429–431).

BN3 Reptilian flight: pterosaurs (chapter 18, pp. 431–436).

BN3 Extinctions and extraterrestrial impacts (chapter 18, table on pp. 432–433).

BN3 Birds (late Cenozoic) (chapter 18, pp. 436–438).

Chapter Eleven – The Carboniferous: Amphibians and Early Mammals; Use of Body Plans for Mammals; Genetic Regulatory Networks (GRNs)

BN3 Monroe W. Strickberger's *Evolution* (see chapter 19).

BN3 "From Water to Air: Amphibians, Reptiles, and Birds" (chapter 18, pp. 409–438).

Review of the evolutionary timing of individual species through all the periods that had life. Use description below to follow evolutionary flow.

BN3 Evolution of mammals. See Fig. 16–18 (cohesive summary of the amniotic tree in one chart). (chapter 18, pp.422-438)

Amniotes initially divided into eight branches. They were synapsids, anapsids, two branches of lepidosaurs, diapsids, archosaurs, and dinosaurs. Each of these, as time went on, divided into two or three branches.

The first branch was synapsids, which ended up producing mammals in the Cenozoic Eon. It did not branch but produced pelycosaurs in the Permian Period and therapsids in the Triassic Period. Then it continued in time until the early Cenozoic Eon and produced mammals.

The second branch, anapsids, produced an interim stage, Pareiasaurus, in the Triassic and then continued unchanged until it reached the Cenozoic Eon when it produced turtles. The next two branches leading out from amniotes made it all the way to the Cenozoic Eon without branching and then produced snakes and lizards. Lepidosaurs, the second branch, proceeded through time from the early Carboniferous to the late Cenozoic without producing any animals. When it reached the late Cenozoic, it produced small dinosaur-like creatures called Sphenacodontia. The fourth branch, diapsids, made its way from the late Carboniferous and progressed as far in time as the middle of the Jurassic and then divided into plesiosaurs, placodontids, and ichthyosaurs. All three branches became evolutionary dead ends in the Jurassic and Cretaceous Eras, long before they made it to the Cenozoic Era.

The archosaur group was a continuation of the original diapsid stem. At the Permian boundary, it branched into three branches. One branch headed out in time and had no offshoots until it reached the middle of the Cenozoic Eon. It continued to the early Cenozoic and produced crocodiles. The second branch started at the Permian boundary and proceeded without branching until it reached a time in the Jurassic/Cretaceous Periods. It produced pterosaurs and dead-ended soon after. That left the diapsid branch bare for a while.

One branch created two branches in the Jurassic. It did no more branching, but in the late Cretaceous, it produced Ornithischian dinosaurs. The ongoing branch remained and, in the mid-Triassic, produced Saurischian dinosaurs and dead-ended before the Cretaceous ended. The second branch produced Ornithischian dinosaurs until it reached the late Cretaceous, when it became extinct.

The branch produced Saurischian dinosaurs until the late Cretaceous when it decreased in numbers and became extinct. The main archosaur branch continued and in the Cretaceous Period produced birds (the order Aves). Soon after, all dinosaurs became extinct, partly due to the asteroid strike in the Gulf of Mexico.

Online References

1. Brian K. Hall, "Evo-Devo: Evolutionary Developmental Mechanisms" (Department of Biology, Dalhousie University, Halifax, Nova Scotia, Canada, 2003). https://pubmed.ncbi.nlm.nih.gov/14756324/.

2. "Body Plan," Wikipedia, the Free Encyclopedia. https://en.wikipedia.org/wiki/Body_plan.

3. Jeffrey P. Tomkins and Jerry Bergman, "Developmental Gene Regulatory Networks—An Insurmountable Impediment to Evolution" (Journal of Creation, 2018), 32(2)96–102. https://creation.com/developmental-gene-regulatory-networks.

4. Jaume Baguna and Jordi Garcia-Fernàndez, "Evo-Devo: The Long and Winding Road" (Department of Genetics, Faculty of Biology, University of Barcelona, Spain, 2003). https://pubmed.ncbi.nlm.nih.gov/14756346/.

This reference is worth reading. Although it involves complex genetics, the overall arguments sound clear. The modern synthesis concept of GRNs is not compatible with the cohesiveness of the existing genetic structure and makes death of the organism more likely.

5. Lee Altenberg, "The Evolution of Evolvability in Genetic Programming" (SFI Working Papers, Santa Fe Institute: Institute of Statistics and Decision Sciences, Duke University, 1994). https://sfi-edu.s3.amazonaws.com/sfi-edu/production/uploads/sfi-com/dev/uploads/filer/99/fa/99fa73e5-7d46-4cd2-9e2d-4c1eed56a25f/94-02-007.pdf.

6. "Neoteny in Humans," Wikipedia, the Free Encyclopedia. https://en.wikipedia.org/wiki/Neoteny_in_humans#:~:text=Neotenic%20features%20of%20the%20head,)%20and%20mandible%20(lower%20jaw.

Chapter 12 – The Permian Triassic Extinction: A Series of Giant Reptiles

BN3 Monroe W. Strickberger's *Evolution* (see chapter 18: "From Water to Air: Amphibians, Reptiles, and Birds").

BN3 The radiation of amniotes (pp. 410–438).

BN3 Early amphibians (pp. 410–414).

BN3 Modern amphibians (Lissamphibia) (pp. 414–416).

BN3 Amphibian tetrapods to amniotes (the Amniotic egg) (pp. 416–419).

BN3 Reptilian evolution (pp. 419–423).

BN3 Early archosaurs and dinosaurs (pp. 424–427).

BN3 Endothermy versus ectothermy (pp. 427–429).

BN3 The late Cretaceous extinctions (pp. 429–433).

BN3 Reptilian flight: pterosaurs (pp. 431–436).

BN3 Birds (pp. 436–438).

During the Permian-Triassic extinction event, also called the Great Dying, 90 to 95 percent of marine species became extinct. Seventy percent of all land creatures also became extinct. It took thirty million years to replace the lost organisms, most of which were insect species. Recurrent outpourings of magma from the Siberian Traps over thousands of years are believed to have been the cause of the destruction. Recovery took thirty million years.

Online References

1. "Siberian Traps," Wikipedia, the Free Encyclopedia. https://en.wikipedia.org/wiki/Siberian_Traps. Probable cause of Permian extinction.

2. "Permian," Wikipedia, the Free Encyclopedia. https://en.wikipedia.org/wiki/Permian.

3. Bob Strauss, "Prehistoric Reptiles That Ruled the Earth Before the Dinosaurs," ThoughtCo, August 27, 2024. https://www.thoughtco.com/reptiles-that-ruled-earth-before-dinosaurs-1093310#:~:text=What%20was%20before%20dinosaurs?,%22)%20that%20preceded%20the%20dinosaurs.

Chapter Thirteen – The Jurassic Period and the Evolution of Birds

BN3 Monroe W. Strickberger's *Evolution* (see chapter 19).

BN3 Evolution of mammals, teeth (pp. 443–461). There are twelve soft-body and hard-body features for mammals. The soft-body features are: mammary glands, live birth, control of body temperature, a diaphragm, a four-chamber heart, and an expanded neocortex with greater intelligence. The hard-body features are: a double occipital condyle that articulates with the first cervical vertebra; a mandible that consists of a single bone called the dentary that articulates with the squamosal bone of the skull; the transformation of quadrate and articular bones of the jaw into incus and malleus ear ossicles; a bony secondary palate separating the nasal passages from the mouth; a single nasal opening in the skull; and a relatively large braincase.

BN3 Teeth (pp. 443–444).

BN3 Jaws and hearing (pp. 444–445).

BN3 Early mammals and early mammal habitats (pp. 446–448).

BN3 Marsupials and placentals (pp. 449–450).

BN3 The Mesozoic experience (pp. 451–452).

BN3 The Cenozoic Era: The age of mammals and the northern continents (pp. 452–454).

BN3 Two island continents: Australia and South America (pp. 454–458).

Author's note on the origin of birds: The creation of birds in a short period of time is addressed as being the work of Elohim. Despite the complex variations, there is evidence that this was not a natural evolutionary process. He created birds with all the features He had so admired in previous universes. He crossbred different species of dinosaurs to produce one that had all the genetic elements that produced the body plans of different species of birds. Scientists believe that this happened naturally, but are surprised at how short a time interval it happened in.

Online References

1. "Synapsida," Wikipedia, the Free Encyclopedia. https://en.wikipedia.org/wiki/Synapsida.

2. "Therapsida," Wikipedia, the Free Encyclopedia. https://en.wikipedia.org/wiki/Therapsida.

3. "Jurassic," Wikipedia, the Free Encyclopedia. https://en.wikipedia.org/wiki/Jurassic.

4. "The Origin of Birds," Berkely University of California, 2025. https://evolution.berkeley.edu/what-are-evograms/the-origin-of-birds/.

5. Yanhong Pan, et al., "The Jehol Biota: Definition and Distribution of Exceptionally Preserved Relics of a Continental Early Cretaceous Ecosystem" (Cretaceous Research, August 2013), 34:30–38. https://www.sciencedirect.com/science/article/abs/pii/S0195667113000566.

Chapter 14 – The Cretaceous Period; Radiation of Flowering Plants; the Cretaceous-Paleogene Extinction

BN3 Monroe W. Strickberger's *Evolution* (see chapter 20, "Primate Evolution and Human Origins").

BN3 Primate classification: lemurs, lorises, tarsiers, platyrrhines, catarrhines, hominoids, gibbons, orangutans, chimpanzees, and gorillas (chapter 20, pp. 463–466).

BN3 Human-ape comparisons (chapter 20, pp. 466–468).

BN3 The fossil record (chapter 20, pp. 468–470).

BN3 The australopithecines (chapter 20, pp. 470–474).

BN3 Bipedalism (chapter 20, pp. 474–477).

BN3 Homo (chapter 20, pp. 477–488).

BN3 Hunting hominids (chapter 20, pp. 489–491).

BN3 Communication and speech (chapter 20, pp. 491–495).

BN3 Language and self-awareness (chapter 20, pp. 495–498).

BN3 Altruism and morality (chapter 20, pp. 498–499, 506).

BN3 Evolution of the human brain (chapter 20, pp. 500–505).

Online References

1. "The Chicxulub Impact Event," the Lunar and Planetary Institute. https://www.lpi.usra.edu/science/kring/Chicxulub/students-teachers/.

2. Richard J. A. Buggs, "The Origin of Darwin's 'Abominable Mystery'" (American Journal of Botany, January 22, 2021), 108(1):22–36. https://bsapubs.onlinelibrary.wiley.com/doi/10.1002/ajb2.1592.

3. "Has Darwin's Abominable Mystery Been Solved or Not?" StackExchange.com question board, accessed October 13, 2025. https://biology.stackexchange.com/questions/111661/has-darwins-abominable-mystery-been-solved-or-not.

   Solution to Darwin's angiosperm display discussed.

Chapter Fifteen – The Cenozoic Eon: Evolution of Primates; Tracking the Flow of the Creation of Animals

BN3 Monroe W. Strickberger's *Evolution* (see chapter 20, "Primate Evolution and Human Origins").

BN3 Primate classification: lemurs, lorises, tarsiers, platyrrhines, catarrhines, hominoids, gibbons, orangutans, chimpanzees, and gorillas (pp. 463–466).

BN3 Human-ape comparisons (pp. 466–468).

BN3 The fossil record (chapter 20, pp. 468–470).

BN3 The australopithecines (chapter 20, pp. 470–474).

BN3 Bipedalism (chapter 20, pp. 474–477).

BN3 Homo (chapter 20, pp. 477–488).

BN3 Hunting hominids (chapter 20, pp. 489–491).

BN3 Communication and speech (chapter 20, pp. 491–495).

BN3 Language and self-awareness (chapter 20, pp. 495–498).

BN3 Altruism and morality (chapter 20, pp. 498–499, 506).

BN3 Evolution of the human brain (chapter 20, pp. 500–505).

Online References

1. Carol Ward and Ashley Hammond, "Australopithecus and Kin" (Nature Education Knowledge, 2016), *Nature Education Knowledge* 7(3):1. https://www.nature.com/scitable/knowledge/library/australopithecus-and-kin-145077614/.

2. "Timeline of Human Evolution," Wikipedia, the Free Encyclopedia. https://en.wikipedia.org/wiki/Timeline_of_human_evolution.

3. Sarah Malmquist and Kristina Prescott, "The Evolution of Primates," in *Human Biology* (University of Minnesota, 2023), accessed October 13, 2025. https://open.lib.umn.edu/humanbiology/chapter/1-7-the-evolution-of-primates/.

Track the flow of the great apes during their known dates of divergence with other primates.

Chapter Sixteen – The Missing African Apes

Online References

1. Marjorie Hecht, "The Mystery of the Missing Apes Who Came Before Humans," Wiki Observatory, June 8, 2024. https://observatory.wiki/The_Mystery_of_the_Missing_Apes_Who_Came_Before_Humans.

2. Tim D. White, et al., "Neither Chimpanzee nor Human, *Ardipithecus* Reveals the Surprising Ancestry of Both," Proc Natl Acad Sci USA, April 20, 2015. https://pmc.ncbi.nlm.nih.gov/articles/PMC4413341/.

Discusses parts played by evolving Ardipithecus, australopiths, and existing apes in reproducing the lost African Great Apes and in the formation of Homo.

Chapter Seventeen – The Bipedal Apes: Ardipithecus and Australopithecines

BN6 Roger Levin and Robert Foley's *Principles of Human Evolution.*

BN6 The apelike hominins (chapter 10, pp. 255–283).

BN6 The australopithecines in outline (chapter 10, pp. 255–259).

BN6 The beginnings of Australopithecus (chapter 10, pp. 259–261).

BN6 The Australopithecus afarensis story (chapter 10, pp. 261–264).

BN6 Anatomy of Au. afarensis (chapter 10, pp. 265–271).

BN6 Australopithecine diversity (chapter 10, pp. 272–273).

BN6 Australopithecine biology (chapter 10, pp. 274–275).

BN6 Australopithecine anatomy (chapter 10, pp. 275–279).

BN6 Australopithecus—a tool maker? (chapter 10, pp. 279–280).

BN6 Relationship between robust and gracile australopithecines (chapter 10, p. 280).

BN6 Paleoenvironments of the early hominins (chapter 10, pp. 280–283).

Online References

1. Tim D. White, et al., "Neither Chimpanzee nor Human, *Ardipithecus* Reveals the Surprising Ancestry of Both," Proc Natl Acad Sci USA, April 20, 2015. https://pmc.ncbi.nlm.nih.gov/articles/PMC4413341/.

Discusses parts played by evolving Ardipithecus, australopiths, and existing apes in reproducing the lost African Great Apes and in the formation of Homo.

2. Ewen Calloway, "Oldest Genetic Data from a Human Relative Found in 2-Million-Year-Old Teeth," NatureNews.com, July 10, 2023.
https://www.nature.com/articles/d41586-023-02242-z.

3. "Taung Child," Britannica, 2025.
https://www.britannica.com/topic/Taung-child.

Chapter Eighteen – Bipedal Apes and the Fossil Trail

- The bipedal apes: A chronology of their short existence and discussion of the reasons why:
- Ar. kadabba – 5.8 mya to 5.2 mya. Not clear how long they persisted. May not be a hominin.
- Ar. ramidus – 4.4 mya. No information on when they became extinct. Same species dated to be 4.4 to 5.8 mya. Origin unclear but has numerous fossils.
- Au. anamensis – 4.1 mya. Disappeared at 3.9 mya. Interval present 0.2 ky (kiloyear). Region Kenya.
- Au. afarensis – 3.7 mya to 3.0 mya. Interval present 0.7 mya. Region Tanzania.

- Taung Child – 3 mya. First fossil found was the skull of a three-year-old.
- Au. bahrelghazali – probably a sub-species of Au. afarensis.
- Au. africanus – 3.0 mya to 2.1 mya. Interval 0.9 mya. Taung, Sterkfontein, Makapansgat, South Africa.
- Paranthropus aethiopicus – 2.7 mya to 0.3 mya. Interval 2.4 mya.
- Au. garhi – Approximately 2.5 mya. Bouri, Ethiopia.
- Paranthropus boisei – 2.3 mya to 1.4 mya. East Africa and South Africa. Interval 0.9 mya.
- Au. sediba – 1.95 mya to 1.78 mya. Malapa, South Africa. Interval 0.17 mya.
- Paranthropus robustus – 1.8 mya to 1.5 mya. South Africa in three areas. Interval 0.3 mya.

Online References

1. "Hominid Species," Fossil Hominids FAQ, April 30, 2010. https://www.talkorigins.org/faqs/homs/species.html.

2. Ian Tattersall, "Homo Sapiens," Britannica, June 16, 2025. https://www.britannica.com/topic/Homo-sapiens.

Chapter Nineteen – Elohim Decides to Create Body Plans for Early Homo Species; H. Habilis, H. Rudolfensis, and Homo Ergaster; He Adds a Female as a Mate for Ergaster

This is where Elohim took over the introduction of Homo and his fellow humans. Since little definitive evidence is available as to how it all came about, paleoanthropologists and archeologists have multiple hypotheses and multiple evolutionary trees leading to the introduction of early Homo. It is interesting, but there is much controversy and there are no cohesive decisions on data. Elohim was closely involved in the final stages of the rise of Homo sapiens sapiens from early archaic Homo sapiens and other interim stages that occurred. He did not unduly influence the process but had the help of H. erectus, and the two H. heidelbergensis leaders interacting with AMH as they traveled and blended with H. neanderthalensis in Eurasia. It took a while, but gradually, the final features of H. sapiens began to be the prevalent morphology.

Chapter Twenty – Homo Ergaster to Homo Erectus Grade

Homo ergaster's last stand: This chapter covers the life history of the species as determined by fossils of Homo ergaster to Homo erectus. Erectus's disappearance for three hundred thousand years after doing his job working with AMH is now explained. In the 1990s, a group of skulls were found in the Ngandong area in Java, in debris backed up in a curve in the Solo River. Until recently, multiple attempts to date the skull caps have been inaccurate due to geologic issues in the area where they were found. They have now been identified as being H. erectus for certain, with the exception that they were larger than previous ones. Dating them has been difficult because of the terrain that the river runs through.

The usual radioactive materials have not yielded appropriate results when performed on hog bones lying close to the skulls in the same layer as the H. erectus fossils. In a recent study, they were treated with thermoluminescence testing and given an age of 117,000 to 108,000 years ago.

Detailed stratigraphic study was implemented to make sure that the layer represented the same strata as the one where the bones were found and it is supposed to be a result that is close to reality.

Online References

1. Barbara Helm Welker, *The History of Our Tribe: Hominini* (OpenSuny Textbooks, 2017), accessed October 13, 2025. https://milnepublishing.geneseo.edu/the-history-of-our-tribe-Hominini/chapter/the-erectus-grade/.

2. Lisa Henry, "*Home Erectus,* Our Ancient Ancestor," Natural History Museum, accessed October 13, 2025. https://www.nhm.ac.uk/discover/homo-erectus-our-ancient-ancestor.html.

Chapter Twenty-One – Evolution of Homo Neanderthalensis and Anatomically Modern Humans (AMH) from an Unknown Ancestor in Spain

BN6 Roger Levin and Robert Foley's *Principles of Human Evolution.*

BN6 The origin of modern humans: Background and fossil evidence (chapter 14, pp. 365–401).

BN6 A note on species and names (chapter 14, pp. 366–372).

BN6 Competing hypotheses for modern human origins; predictions of different hypotheses (chapter 14, pp. 372–374).

BN6 Chronological evidence: Where were the earliest anatomically modern humans? (chapter 14, pp. 375–376).

BN6 The question of regional continuity: Southeast Asia, Australasia, East Asia, Middle East, Europe, and Africa (chapter 14, pp. 377–391).

BN6 The fossil evidence as a whole (chapter 14; pp. 391–394).

BN6 The place of Neanderthals in human evolution (chapter 14, pp. 394–398).

Online References

1. "Archaeological Site at Atapuerca," Wikipedia, the Free Encyclopedia.

https://en.wikipedia.org/wiki/Archaeological_site_of_Atapuerca.

Oldest bones found so far in Europe.

2. Alison George, "Neanderthals Never Lived in Africa, but Their Genes Got There Anyway," NewScientist, January 30, 2020. https://www.newscientist.com/article/2231991-neanderthals-never-lived-in-africa-but-their-genes -got-there-anyway/.

3. "Researchers Determine Age for Last Known Settlement by a Direct Ancestor to Modern Humans," Science Daily. University of Iowa, December 18, 2019. https://www.sciencedaily.com/releases/2019/12/191218153527.htm#google_vignette.

Chapter Twenty-Two – Shanidar and Nandi

Online References

1. "Shanidar I," Smithsonian Institution, 2025. https://humanorigins.si.edu/evidence/human-fossils/fossils/shanidar-1.

Chapter Twenty-Three – Eruption of Toba Volcano in Sumatra; Archaic Homo Sapiens Travel to Asia

Personalization is imagined as are fine details of the journey. Such journeys have been projected to have happened by paleoanthropologists. This applies to all chapters that are personalized to Elohim's perspective.

Online References

1. "Youngest Toba Eruption," Wikipedia, the Free Encyclopedia. https://en.wikipedia.org/wiki/Youngest_Toba_eruption.

2. Alison George, "Neanderthals Never Lived in Africa, but Their Genes Got There Anyway," NewScientist, January 30, 2020. https://www.newscientist.com/article/2231991-neanderthals-never-lived-in-africa-but-their-genes -got-there-anyway/.

3. Jane Qiu and Nature Magazine, "How China Is Rewriting the Book on Human Origins," Scientific American, July 13, 2016. https://www.scientificamerican.com/article/how-china-is-rewriting-the-book-on-human-origins/.

4. "Timeline of Human Evolution," Wikipedia, the Free Encyclopedia. https://en.wikipedia.org/wiki/Timeline_of_human_evolution.

Chapter Twenty-Four – The Garden of Eden: Adam and Eve

BN1 The Holy Bible. The fall of man. See Genesis 3:1–24.

Chapter Twenty-Five – Satan's Initial Victory over Man; Elohim Introduces Speech as a Tool for Communication to Homo Sapiens and H. Heidelbergensis

BN6 Roger Levin and Robert Foley's *Principles of Human Evolution.*

BN6 The origin of modern humans: genetic evidence (pp. 400–421).

BN6 The impact of molecular evolutionary genetics (pp. 401–409).

BN6 The mitochondrial Eve story (chapter 5, pp. 417–421).

Elohim gives the gift of speech to the Homo sapiens to enable them to communicate as they adjust to their new life. This chapter is a personalized account of Homo sapiens exploring the world. In keeping with the behavior of Homo at this stage, many routes by which AMH left Africa have been suggested. Once in Europe and Asia, most AMH stayed there but some returned to Africa, bringing back genes of other species.

Online References

1. Ian Tattersall, "Homo sapiens," Britannica, June 16, 2025. https://www.britannica.com/topic/Homo-sapiens.

   Includes sections on bodily structure and evolution.

2. "Homo heidelbergensis," Smithsonian Institution, January 3, 2024. https://humanorigins.si.edu/evidence/human-fossils/species/homo-heidelbergensis.

3. "Timeline of Human Evolution," Wikipedia, the Free Encyclopedia. https://en.wikipedia.org/wiki/Timeline_of_human_evolution.

Chapter Twenty-Six – Homo Sapiens Sapiens Occupy the Earth; the Beginning of the Journey from Eden and out of Africa

BN6 Roger Levin and Robert Foley's *Principles of Human Evolution.*

BN6 The origin of modern humans: genetic evidence (pp. 400–421).

BN6 A note on species and names (chapter 14, pp. 366–372).

BN6 Competing hypotheses for modern human origins; predictions of different hypotheses (chapter 14, pp. 372–374).

BN6 Predictions of different hypotheses (chapter 14, p. 374).

BN6 Chronological evidence: Where were the earliest anatomically modern humans? (chapter 14, pp. 375–376).

BN6 The question of regional continuity: Southeast Asia, Australasia, East Asia, Middle East, Europe, and Africa (chapter 14, pp. 377–391).

BN6 The fossil evidence as a whole (chapter 14, pp. 391–394).

The fossil evidence as a whole is closer to "out of Africa" as opposed to the multiregional one. Besides the protagonist for this choice, there are individual paleontologists who suggest a single origin model. Currently the first model is more popular. This suggests more than one dispersal. The topic remains an actively discussed one. The relationship between Homo, which originated in Africa, and the Neanderthal, which originated in Eurasia, is still being studied. Recent findings at the site in Antequera, Spain, have led to rearranging the genealogy of the four most crucial species: Homo sapiens, Homo heidelbergensis, Homo neanderthalensis, and the Denisovans.

The next two chapters delve into the part played by genetics and are more complex but very interesting to read. One issue is all but settled when taking into account all the genetic and historical data. The final human product, AMH to Homo sapiens, is thought to have evolved in Africa. Many gaps and unsettled questions still exist, and new specimens of fossils continue to be dug up, so things could change.

The imagined story, supported by plausible events, is supported by evidence that Homo sapiens traveled in groups from Africa and Asia via a variety of routes and then returned to Africa. It explains the eventual findings of only Homo sapiens and a branch of Homo heidelbergensis merging somehow to give rise to AMH and from there very quickly to only H. sapiens sapiens. The shifting of activity from Eurasia, which has genetic evidence because of cool summers that preserve DNA, makes it difficult to compare it with the African genetic data. Since he emerged, the conversion of archaic Homo sapiens to Homo sapiens sapiens has occurred. The mechanism by which the three groups blended or became extinct to produce only one group, Homo s. sapiens, is not clear. Many techniques, like molecular clock processing, the use of mitochondrial DNA, which can provide population info, and fossils, have added to the whole picture. Fossils are especially useful to determine the entry of Homo to Australia and other islands. The period during which many changes occurred included the time when Elohim gave all the Homo species the anatomic and cerebral changes that would allow the development of speaking and hearing the spoken word and for language to develop to communicate. As they traveled, they developed an ability to communicate and slowly a common language evolved. They also upgraded their tools, going from Oldowan to Mousterian. At the end of their walk, on Elohim's direction they ended up in Afghanistan.

Chapter Twenty-Seven – Karstagians: Remnants of the Population That Lived Around Eden

Elohim returns after one hundred years to the small settlement in Afghanistan. He is impressed with His group of farmers in the progress they have made. They are now farmers and not hunter-gatherers. Beelzebub, envious of Elohim, decides to send trouble to the settlement. They are forced to leave, and the Nephilim help them with the move. The chapter goes on to describe their forming into three groups and heading in three different directions to find their own safe place. They are in constant contact with Elohim. They each settle in a different farming area which prospered, giving rise to the Yellow Valley civilization in China, the Sumerian civilization in the valleys of the Euphrates, the Tigris Rivers in what today is Iraq, the Tehuacana Valley civilization in Mexico, and the Indus Valley civilization on the shores of the Indus River. They settled in the two deltas of the estuaries of the Tigris and the Euphrates Rivers in the area that is now Iraq. The cities built in these cradles of civilization still have structures that you can see today if you travel to them.

Chapter Twenty-Eight – Four Thousand Years Ago: The Cradle of Civilization

**Note: There are no Endnotes or Sources for this chapter.*

Chapter Twenty-Nine – The Denisovans

Online References

1. "Denisovan," Wikipedia, the Free Encyclopedia. https://en.wikipedia.org/wiki/Denisovan.

Chapter Thirty – The Indus Valley Civilization: The Great Flood

Online References

1. "Indus Valley Civilization," Wikipedia, the Free Encyclopedia. https://en.wikipedia.org/wiki/Indus_Valley_Civilisation.

2. William Ryan and Walter Pitman, *Noah's Flood: The New Scientific Discoveries About the Event That Changed History* (Simon & Schuster, 2000).

3. Alfred Hamori, "The Origin of the Sumerians and the Great Flood," excerpt from Starinthestone.wordpress.com, accessed October 13, 2025. https://starinthestone.wordpress.com/2012/07/01/1164/.

   He suggests that the polar shift in 12,000 BCE caused a major climatic change in the steppes of Siberia which progressed from a moderate temperate to an arctic climate and a prolonged winter.

www.ingramcontent.com/pod-product-compliance
Lightning Source LLC
LaVergne TN
LVHW052223200926
848481LV00001B/191